Multiphoton Ionization of Atoms

Contributors

P. Agostini

S. L. Chin

Michèle Crance

N. B. Delone

F. Fabre

D. Feldmann

Y. Gontier

C. Jung

H.-J. Krautwald

P. Lambropoulos

G. Leuchs

G. Mainfray

C. Manus

G. Petite

V. V. Suran

M. Trahin

H. Walther

A. Weingartshofer

K. H. Welge

P. Zoller

B. A. Zon

Multiphoton Ionization of Atoms

Edited by

S. L. CHIN

Laboratoire de Recherches en Optique et Laser
Département de Physique
Faculté des Sciences et de Génie
Université Laval
Québec, Canada

P. LAMBROPOULOS

Department of Physics
University of Southern California, University Park
Los Angeles, California
and
University of Crete
Iraklion, Crete
Greece

 1984

ACADEMIC PRESS

(Harcourt Brace Jovanovich, Publishers)

Toronto   Orlando   San Diego   San Francisco   New York

London   Montreal   Sydney   Tokyo   São Paulo

7183 - 6895

PHYSICS

ACADEMIC PRESS CANADA
55 Barber Greene Road, Don Mills, Ontario   M3C 2A1

United States Edition published by ACADEMIC PRESS, INC.
Orlando, Florida 32887

United Kingdom Edition published by ACADEMIC PRESS, INC. (LONDON) LTD.
24/28 Oval Road, London NW1   7DX

Library of Congress Cataloging in Publication Data
Main entry under title:

Multiphoton ionization of atoms.

   1. Photoionization. 2. Multiphoton processes.
I. Chin, S. L. II. Lambropoulos, Peter.
QC702.M84   1984   539.7'54   83-9996
ISBN 0-12-172780-7

Canadian Cataloguing in Publication Data

Main entry under title:
Multiphoton ionization of atoms

Bibliography
Includes index.
ISBN 0-12-172780-7

   1. Ionization. 2. Multiphoton processes. I. Chin,
S. L. II. Lambropoulos, Peter.

QC702.M84   539.7'54   C83-098663-4

84 85 86 87   9 8 7 6 5 4 3 2 1

Contents

1 Introduction   1

S. L. Chin

2 "Normal" Multiphoton Ionization of Atoms (Experimental)   7

G. Mainfray and C. Manus

3 Theory of Multiphoton Ionization of Atoms   35

Y. Gontier and M. Trahin

List of Contributors

Numbers in parentheses indicate the pages on which the authors' contributions begin.

P. Agostini (133), Service de Physique des Atomes et des Surfaces, Centre d'Etudes Nucléaires de Saclay, 91191 Gif-sur-Yvette Cédex, France

S. L. Chin (1), Laboratoire de Recherches en Optique et Laser, Département de Physique, Faculté des Sciences et de Génie, Université Laval, Québec, Canada G1K 7P4

Michèle Crance (65), Laboratoire Aimé Cotton, CNRS II, 91405 Orsay, France

N. B. Delone (235), General Physics Institute, Academy of Sciences of the USSR, Moscow, USSR 117924

F. Fabre (133), Service de Physique des Atomes et des Surfaces, Centre d'Etudes Nucléaires de Saclay, 91191 Gif-sur-Yvette Cédex, France

D. Feldmann (223), Fakultät für Physik, Universität Bielefeld, D4800 Bielefeld 1, Federal Republic of Germany

Y. Gontier (35), Service de Physique des Atomes et des Surfaces, Centre d'Etudes Nucléaires de Saclay, 91191 Gif-sur-Yvette Cédex, France

C. Jung (155), Fachbereich Physik, Universität Kaiserslautern, 675 Kaiserslautern, Federal Republic of Germany

H.-J. Krautwald (223), Fakultät für Physik, Universität Bielefeld, D4800 Bielefeld 1, Federal Republic of Germany

P. Lambropoulos (189), Department of Physics, University of Southern California, University Park, Los Angeles, California 90007, and University of Crete, Iraklion, Crete, Greece

*G. Leuchs** (109), Sektion Physik, Universität München, 8046 Garching, Federal Republic of Germany

*Present address: Joint Institute for Laboratory Astrophysics, University of Colorado, Boulder, Colorado 80309.

G. *Mainfray* (7), Service de Physique des Atomes et des Surfaces, Centre
    d'Etudes Nucléaires de Saclay, 91191 Gif-sur-Yvette Cédex, France
C. *Manus* (7), Service de Physique des Atomes et des Surfaces, Centre
    d'Etudes Nucléaires de Saclay, 91191 Gif-sur-Yvette Cédex, France
G. *Petite* (133), Service de Physique des Atomes et des Surfaces, Centre
    d'Etudes Nucléaires de Saclay, 91191 Gif-sur-Yvette Cédex, France
V. V. *Suran* (235), Uzhgorod State University, Uzhgorod, USSR 294000
M. *Trahin* (35), Service de Physique des Atomes et des Surfaces, Centre
    d'Etudes Nucléaires de Saclay, 91191 Gif-sur-Yvette Cédex, France
H. *Walther* (109), Sektion Physik, Universität München, and Max-
    Planck-Institut für Quantenoptik, 8046 Garching, Federal Republic
    of Germany
A. *Weingartshofer* (155), Department of Physics, St. Francis Xavier Uni-
    versity, Antigonish, Nova Scotia, Canada B2G 1C0
K. H. *Welge* (223), Facultät für Physik, Universität Bielefeld, D4800
    Bielefeld 1, Federal Republic of Germany
P. *Zoller* (189), Institute für Theoretische Physik, Universität Innsbruck,
    Innrain, Innsbruck, Austria
B. A. *Zon* (235), Department of Physics, Voronezh State University,
    Voronezh, USSR 394693

Preface

Multiphoton ionization of atoms, molecules, and surfaces has evolved into a large field of research encompassing a wide spectrum of problems from basic to very applied. The ionization of atoms was the part of the field that began developing first and as a result is at a somewhat more advanced state at this time. Having passed the stage of qualitative understanding, the emphasis from now on will be on quantitative comparisons between theory and experiment. One aspect of such comparisons is of a spectroscopic character in that it aims at the understanding of atomic structure—especially excited states—which is impossible or extremely difficult to obtain otherwise. Another aspect has to do with the nonlinear character of the interaction itself, which brings in new physics. The large intensity within narrow bandwidths leads to significant saturation of resonant transitions; the statistical properties (coherence) of the laser radiation have a dramatic effect on the nonresonant as well as the resonant multiphoton processes. The combination of resonance with laser temporal fluctuations has in the past four years or so led to a number of surprises, especially in the context of ac Stark splitting. Thus, intermediate resonances, saturation, ac Stark shifts and splitting, and field-fluctuation effects are pivotal aspects of multiphoton processes.

Although several review articles have appeared over the past few years, most of the material constituting the present body of knowledge remains scattered in the original papers or at most summarized in reviews of specific aspects. The intention of this volume is to provide the reader with a pedagogical review of essentially the whole subfield of multiphoton ionization of atoms in the form of a collection of chapters written by active participants in the development of the particular area. The multiplicity of authors inevitably entails some sacrifice of coherence in return for the authenticity gained by having contributions from leading researchers in the field. The result cannot be as methodical a book as a textbook, but it can nevertheless introduce the reader to the field and provide an exposition of its present status.

Despite its relatively advanced stage, multiphoton ionization of atoms is in a state of flux. In a single-electron model, the theory is by and large understood. When more than one electron excitation is involved, however, the field is more or less open. When free–free transitions are involved—either in scattering in the presence of a laser field or in the continuum above the ionization threshold of an atom—much remains to be clarified. Multiple ionization through multiphoton absorption is only beginning to be explored. Parts of this book, therefore, deal with topics of some maturity while others must be viewed as somewhat tentative. Much of the content of the following chapters will nevertheless remain as a permanent part of the body of knowledge in multiphoton processes, not only in atoms but, with minor modifications, in many other contexts of multiphoton absorption by bound electrons.

Multiphoton Ionization of Atoms

1

Introduction

S. L. CHIN

Laboratoire de Recherches en Optique et Laser
Département de Physique
Faculté des Sciences et de Génie
Université Laval
Québec, Canada

The first work on multiphoton processes can be dated as far back as the beginning of quantum theory when Goeppert-Mayer (1931) published a theoretical paper on simultaneous two-photon absorption. Because of the lack of a proper intense monochromatic light source, such an experiment could not be imagined. The advent of the powerful Q-switched ruby laser in the early 1960s and its capability to create breakdown in gases (Meyerand and Haught, 1963, 1964) has sparked a strong interest in explaining the physical mechanism involved. Multiphoton ionization was proposed to be the initiating mechanism that created the first few free electrons, which were then accelerated via the inverse Bremsstrahlung process followed by cascade ionization, i.e., breakdown (Chin, 1970; Morgan, 1975). Almost immediately multiphoton ionization itself became a separate subject of its own because of the challenge of predicting (Keldysh, 1965; Bebb and Gold, 1966), iso-lating, and observing such a highly improbable phenomena. Voronov and Delone (1966) were the first to observe multiphoton ionization of rare-gas atoms followed by the French Saclay group (Agostini *et al.*, 1968). Initial works of different groups (Chin *et al.*, 1969) concentrated mainly on proving the existence of such nonresonant multiphoton phenomena. Soon the question of resonant multiphoton ionization was asked, studied, and understood. The experimental parts of these subjects are reviewed in Chapter 2, while Chapters 3 and 4 give some detailed theories. At the same time, the effects of laser coherence (Chapter 2) and laser polarization (Chapter 5) were studied.

1

Most of the experimental results were rather qualitative because of the difficulty in measuring precisely the laser intensity distribution in the focal volume and the number of ions created. So far only very few experiments have given good quantitative measurement of some multiphoton ionization cross sections (Cervenan *et al.*, 1975; Morellec *et al.*, 1980).

Meanwhile, a different question was posed. What would happen if the laser were very intense? This fundamental question presented a challenge in quantum electrodynamics. Many theories have been proposed, from the early work of Keldysh (1965) to the lastest one of Krstic and Mittleman (1982), all giving different results. Traditionally, one uses the so-called Keldysh-type parameter γ to distinguish the "boundary" between multiphoton ionization ($\gamma \gg 1$) and very intense field ionization or "tunnelling" ($\gamma \ll 1$); here,

$$\gamma \equiv (\omega/eE)(2mI_0)^{1/2}$$

where ω and E are the angular frequency and peak electric field of the laser radiation, respectively; I_0 is the ionization energy of the atom; and e and m are the charge and mass of the electron, respectively. One rare but serious attempt to observe the tunnel ionization was published by Lompré *et al.* (1976). They used a 30-ps Nd:glass laser (1.06 μm) whose peak intensity was $\sim 10^{15}$ W cm^{-2} to ionize rare gases. Under these conditions, $\gamma \sim 0.3$ (not $\gamma \ll 1$). What they observed were purely multiphoton effects (Chapter 2). This could be explained in the following manner. The laser photon energy ($h\nu = 1.17$ eV) was so large that multiphoton ionization would take place in the wings of the laser pulse. Even if there could be some signal due to tunnelling at the peak of the laser pulse, it would be overwhelmed by the multiphoton signal. As of now, there is still no proof that tunnel ionization would occur. An obvious alternative is to use a longer wavelength laser so that multiphoton ionization is almost impossible. Thus if the intensity of such a laser is sufficiently high so that $\gamma \ll 1$, multiphoton ionization would not interfere, and any ionization signal observed in such an experiment could be due to tunnelling. The powerful TEA-CO$_2$ laser is an example ($h\nu = 0.117$ eV). Mainfray (1977) has reported verbally such an attempt with negative result. No details were given.

A very recent experiment in our laboratory (Chin *et al.*, 1983) showed that intense ($\sim 10^{12}$ W cm^{-2}) 4-ns CO$_2$ laser pulses at 10.6 μm have given rise to direct ionization of Kr and Xe atoms. ($\gamma \sim 0.7$) Unfortunately the intensity of the laser was not high enough to give a good intensity dependence of ion signals spanning several orders of magnitude. As such, comparison with existing theories is still not possible. We only know that the slope is very steep around 10^{12} W cm^{-2}. More experiments will be done in the near future using a much better CO$_2$ laser facility.

So far only ions were collected during the experimental work. Electrons emitted were usually assumed to have the lowest possible kinetic energy corresponding to $(khv - I_0)$, where k is the minimum number of photons required to ionize the atom. This assumption did not last long. Soon high-energy electrons were observed to have orginated from the focal volume during multiphoton ionization. (Martin and Mandel, 1976; Hollis, 1978). The origin of these electrons was attributed to the intensity gradient at the focal volume giving rise to the ponderomotive force to accelerate the free electrons created by multiphoton ionization. Electrons with energies up to 1 keV have been observed (Boreham and Luther-Davies, 1979).

Other questions remained. Could the free electrons continue to absorb photons soon after they were excited into the continuum via multiphoton absorption of the minimum number of photons required for ionization? The answer to this so-called above threshold ionization problem is yes (Agostini *et al.*, 1979) and will be discussed in Chapter 5. This phenomenon is still very new and more studies, both theoretical and experimental, are expected to be published in the near future.

Other legitimate questions were asked in relation to acceleration of the free electrons created by multiphoton ionization. For example, how do these electrons absorb entire photons when they were in the continuum? An answer is by way of inverse Bremsstrahlung in which photons were absorbed by an electron colliding with the ion so as to conserve momentum. An un-ambiguous way to observe such a phenomenon is to use an atomic beam crossed by a monoenergetic electron beam and a laser beam. The electrons are expected to both gain and lose energies corresponding to an integral number of photons. Such so-called free–free transitions have indeed been observed by Weingartshofer *et al.* (1977, 1979, 1981, 1983) and will be dis-cussed in detail in Chapter 6. So far only a CO_2 laser has been used in one laboratory and much more work is expected to be done by different groups in the future.

A natural question to be asked after the observation of multiphoton ionization creating singly charged ions is whether multiply charged ions could be created. In the early days even singly charged ions were thought to be too improbable to be observed. It thus became almost unthinkable for multiply charged ions to occur simply because it involved much higher order processes. However, it was later found that observation of such multiply charged ions is still possible. Aleksakhin *et al.* (1979) first report the observa-tion of doubly charged ions from multiphoton ionization of alkaline earth atom. This was soon followed by Chin *et al.* (1980), Feldman *et al.* (1982), and (Feldman and Welge, 1982) who observed the creation of Sr^{2+} and Sr^+. The explanation of the creation of these doubly charged ions is still not clear. It could be due to either stepwise ionization or direct two-electron excitation.

Possible resonance with the low-lying autoionizing states of these atoms may be involved. Chapters 8 and 9 are devoted to the experiments on this phenomenon while Chapter 7 gives a theoretical analysis of the multiphoton interaction with the autoionizing states. It is again expected that more work will be published in the future for a better understanding of this phenomenon.

Very recently, the Saclay group (L'Huillier *et al.*, 1982) has observed multiply charged ions from multiphoton ionization of krypton. Ions of charge up to Kr^{4+} were observed. This question of multiple-charge creation through multiphoton ionization becomes a new challenge in this field.

With the very recent development of femtosecond (10^{-15} s) laser technology, one can now, at least, consider doing some experiments using femtosecond laser pulses to answer the fundamental quantum-mechanical question of instantaneous excitation. This temporal effect is briefly discussed in Chapter 2. (See also Dixit *et al.*, 1980.) An alternate question could be the following. If the spectral width of the coherent femtosecond laser pulse (in the visible, for example) covers most of the radiative transitions of an atom, do we still need to use a very high intensity pulse to do the ionization?

In summary, although much has been done in the past 17 years in the understanding of the fundamentals of multiphoton ionization of atoms, new questions and new challenges such as above-threshold ionization, free–free transition, multiply charged ions creation, resonance with autoionizing states, very intense laser field effect (tunnelling), femtosecond ionization wait to be solved. The field is still as fresh as ever.

REFERENCES

Agostini, P., Barjot, G., Bonnal, J. F., Mainfray, G., and Manus, C. (1968). *IEEE J. Quantum Electron.* **QE-4**, 667.

Agostini, P., Fabre, F., Mainfray, G., and Petite, G. (1979). *Phys. Rev. Lett.* **42**, 1127.

Aleksakhin, I. S., Delone, N. B., Zapesochnyi, I. P., and Suran, V. V. (1979). *Sov. Phys. JETP Engl. Trans.* **49**, 477.

Bebb, B., and Gold, A. (1966). *Phys. Rev.* **143**, 1.

Boreham, B. W., and Luther-Davies, B. (1979). *J. Appl. Phys.* **50**, 2533.

Cervenan, M. R., Chan, R. H., and Isenor, N. R. (1975). *Can. J. Phys.* **53**, 1573.

Chin, S. L. (1970). *Can J. Phys.* **48**, 1314.

Chin, S. L., Isenor, N. R., and Young, M. (1969). *Phys. Rev.* **188**, 7.

Chin, S. L., von Hellfeld, A., Krautwald, J., Feldman, D., and Weldge, K. H. (1980). *2nd Int. Conf. Multiphoton Processes, Budapest, 1980.* (Abstr.)

Chin, S. L., Farkas, Gy, and Yergeau, F. (1983). *J. Phys. B.: At. Mol. Phys.* **16**, L223.

Dixit, S. N., Georges, A. T., Lambropoulos, P., and Zoller, P. (1980). *J. Phys. B.* **13**, L157.

Feldmann, D., and Welge, K. H. (1982). *J. Phys. B.* **15**, 1651.

Feldman, D., Krautwald, J., Chin, S. L., von Hellfeld, A., and Welge, K. H. (1982). *J. Phys. B* **15**, 1663.

Goeppert-Mayer, M. (1931). *Ann. Phys. (Leipzig)* **9**, 273.

Hollis, M. J. (1978). *Opt. Commun.* **25**, 395.

Keldysh, L. V. (1965). *Sov. Phys. JETP, Engl. Trans.* **20**, 1307.

Krstic, P., and Mittleman, M. H. (1982). *Phys. Rev. A* **25**, 1568.

L'Huillier, A., Lompré, L. A., Mainfray, G., and Manus, C. (1982). *Phys. Rev. Lett.* **48**, 1814; also *J. Phys. B: At. Mol. Phys.* **16**, 1363 (1983).

Lompré, L. A., Mainfray, G. Manus, C., Repoux, S., and Thebault, J. (1976). *Phys. Rev. Lett.* **36**, 949.

Mainfray, G. (1977). Invited talk at the *Multiphoton Processes, Int. Conf. 1977*.

Martin, E. A., and Mandel, L. A. (1976). *Appl. Opt.* **15**, 2378.

Meyerand, R. G., and Haught, A. F. (1963). *Phys. Rev. Lett.* **11**, 401; also ibid. (1964), **13**, 7.

Morellec, J., Normand, D., Mainfray, G., and Manus, C. (1980). *Phys. Rev. Lett.* **44**, 1394.

Morgan, C. G. (1975). *Rep. Prog. Phys.* **38**, 621.

Voronov, C. S., and Delone, N. B. (1966). *Sov. Phys. JETP Engl. Trans.* **23**, 54.

Weingartshofer, A., Holmes, J. K, Caudle, G., Clarke, E. M., and Krüger, H. (1977). *Phys. Rev. Lett.* **39**, 269–270.

Weingartshofer, A., Clarke, E. M., Holmes, J. K., and Jung, C. (1979). *Phys. Rev. A* **19**, 2371–2376.

Weingartshofer, A., and Jung, C. (1979). *Phys. Canada* **35**, 119–124.

Weingartshofer, A., Holmes, J. K., and Sabbagh, J. (1981). *Laser Spectroscopy V, Proc. Int. Conf., 5th Jasper, Alberta, Canada*, pp. 247–250. Springer-Verlag, Berlin and New York.

Weingartshofer, A., Holmes, J. K., Sabbagh, J., and Chin, S. L. (1983). *J. Phys. B: Atom. Mol. Phys.* **16**, 1805.

2

"Normal" Multiphoton Ionization of Atoms (Experimental)

G. MAINFRAY AND C. MANUS

Service de Physique des Atomes et des Surfaces
Centre d'Etudes Nucléaires de Saclay
Gif-sur-Yvette, France

I. INTRODUCTION

Multiphoton ionization of atoms is a typical example of one of the new fields of investigation in atomic physics that have been opened by lasers. Such experiments had to be limited to qualitative investigations until it became possible to control the parameters of powerful pulsed lasers. It is only over the past six years that profitable quantitative comparisons between theory and experiment have been made thanks to advances in sophisticated theoretical treatments, as well as to the possibility of conducting very accurate experiments through a better control of all the parameters of powerful pulsed lasers. Since the physics of the multiphoton ionization of atoms is now well understood, it is the right time to present a survey of this mature field.

This chapter gives a survey of the main experimental results on multiphoton ionization of atoms obtained in the Atomic Physics Service at Saclay. These results were obtained by a working group whose members are C. Manus, G. Mainfray, J. Morellec, L-A. Lompré, Y. Gontier, M. Trahin, P. Agostini, G. Petite, D. Normand, and J. Thebault. This working group associated themselves on many occasions with students preparing Ph.D.'s.

It is essential to begin with the basic physics involved. An atom with an ionization energy E_i can be ionized by photons with an energy $h\nu$ much less than E_i if the photon flux is strong enough, which, from a practical point of view, can only be achieved with laser radiation. In this case the atom has to absorb several photons from the laser radiation in order to be ionized. This can be done using two different methods with two very different intensity ranges.

Figure 1 shows schematically the first method, using as an example the ionization of an atom through the absorption of three photons of different energies E_1, E_2, and E_3. Each absorbed photon matches the energy difference between two atomic states. For the different jumps, each photon has a platform to step on, which enormously facilitates the transition. The lifetime of the intermediate atomic states is typically 10^{-8} s. This multistep ionization process can be performed using dye lasers delivering different laser frequencies with an intensity of about 1 kW cm^{-2}.

The second method, designated multiphoton ionization, requires a much higher laser intensity and can be performed with a single laser. Figure 2a shows schematically the 4-photon ionization of an atom. The vertical arrows indicate the photons absorbed in the 4-photon transition from the ground state to the continuum. One of the most essential features of a multiphoton

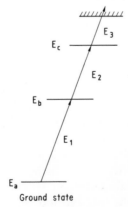

Fig. 1. Schematic representation of the three-step ionization of an atom.

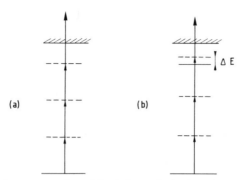

Fig. 2. Schematic representation of the 4-photon ionization of an atom: (a) nonresonant process; (b) quasi-resonant process where ΔE is the energy defect between the energy of three photons and the energy of the closest allowed atomic state.

absorption process is that it occurs through laser-induced virtual states, which are not eigenstates of the atom. In principle, such a multiphoton ionization process does not require any intermediate atomic state. The laser-induced virtual states related to the photon energy and its harmonics act as atomic states; the corresponding lifetimes are, however, much shorter. We may roughly regard the atom as spending a time τ in a laser-induced virtual excited state. This time τ is of the order of one optical cycle, typically 10^{-15} s. Consequently, the absorption of photons through laser-induced virtual states must occur within a time $<10^{-15}$ s. Therefore the photon flux has to be strong enough for having a large number of photons within 10^{-15} s. We thus understand why multiphoton ionization processes can only be achieved with an intense laser radiation.

We now consider a more realistic situation because the last but one photon is absorbed in the dense part of the atomic energy spectrum. When an atomic state is located not too far from a laser-induced virtual state, the aforementioned time τ can be determined by $1/\Delta E$, where ΔE is the energy defect as shown in Fig. 2b, i.e., $\tau = 3 \times 10^{-11}$ s for $\Delta E = 1$ cm^{-1}. Such a quasi-resonant process requires a lower laser intensity. Furthermore, the resonant multiphoton ionization of an atom, corresponding to $\Delta E = 0$, leads to very interesting effects which will be considered in detail in Section III.

II. NONRESONANT MULTIPHOTON IONIZATION OF ATOMS

Nonresonant multiphoton ionization of atoms is the subject of one of the chapters of Volume 18 of the *Advances in Atomic and Molecular Physics* series (Morellec *et al.*, 1982). Consequently, we will only briefly survey this topic.

A. Nonresonant Multiphoton Ionization Induced by Coherent Laser Pulses

The N-photon ionization rate W is given by $W = \sigma_N I^N$, where σ_N is the generalized N-photon ionization cross section. W is expressed in reciprocal seconds, σ_N is expressed in cm^{2N} s^{N-1} units and the laser intensity I in photons per square centimeter per second. Multiphoton ionization cross sections have mainly been measured for alkaline atoms and rare gases with currently available solid-state laser, at a few selected wavelengths, and with laser intensities ranging from 10^9 W cm^{-2} (2-photon ionization of alkaline atoms) to 10^{15} W cm^{-2} (22-photon ionization of helium). Since 1975 it has been possible to obtain accurated σ_N values with the availability of lasers which have good spatial and temporal coherence.

Before giving absolute values for the nonresonant multiphoton ionization cross sections of different atoms, we will present a very simple argument that leads to an order of magnitude estimate of the N-photon ionization cross section. Let us consider the simplest case, the 2-photon ionization of an atom with a laser frequency ω and an intensity I (Gold, 1969). If an energy-conserving first-order transition were possible, the 1-photon transition would take place at a rate $w = \sigma_1 I$, where σ_1, the 1-photon absorption cross section, is typically 10^{-17} cm^2. A second photon can be absorbed if it is incident within the time τ, which is of the order of ω^{-1}, i.e., 10^{-15} s. Again the rate of the second event is $\sigma_1 I$ so that the overall rate for the 2-photon ionization will be

$$w \simeq \sigma_1 I \omega^{-1} \sigma_1 I \qquad (2.1)$$

Therefore $\sigma_2 = w/I^2 \simeq 10^{-49}$ cm^4 s. If there is an atomic state not too far from the laser-induced virtual state, we must replace $\tau = \omega^{-1}$ by $1/\Delta E$, where ΔE is the energy defect as shown in Fig. 2b.

Even though the argument leading to this crude estimate should not be taken literally, the numerical value obtained is in good agreement with experimental and theoretical data. For example, the 2-photon ionization cross section of cesium atoms at 528 nm has been measured to be $\sigma_2 = (6.7 \pm 1.9) \times 10^{-50}$ cm^4 s (Normand and Morellec, 1980). It is in good agreement with different calculations (Bebb, 1966; Teague et al., 1976; Crance and Aymar, 1980; Rachman et al., 1979) as shown in Table I.

As far as 3-photon ionization processes are concerned, let us first return to the simple argument which led to an order of magnitude estimate for σ_2. The overall rate for the 3-photon ionization is

$$w = \sigma_1 I \tau \sigma_1 I \tau \sigma_1 I \qquad (2.2)$$

Therefore,

$$\sigma_3 = w/I^3 \simeq 5 \times 10^{-81} \ cm^6 \ s^2 \qquad (2.3)$$

Table I

2-Photon Ionization Cross Section of Cs

	Reference
Measurement	
$(6.7 \pm 1.9) \times 10^{-50}$ cm^4 s at $\lambda = 528$ nm	Normand and Morellec (1980)
Calculations	
9×10^{-50} cm^4 s	Bebb (1966)
1.2×10^{-49} cm^4 s	Teague *et al.* (1976)
1.2×10^{-49} cm^4 s	Crance and Aymar (1980)
6.6×10^{-50} cm^4 s	Rachman *et al.* (1979)

This estimate gives also the right order of magnitude with experimental and theoretical data on 3-photon ionization cross sections of triplet and singlet metastable helium atoms which are respectively, at 6943.5 nm, $(3.0 \pm 2.2) \times 10^{-81}$ cm^6 s^2 and $(3.3 \pm 1.9) \times 10^{-80}$ cm^6 s^2 (Lompré *et al.*, 1980a). The higher value obtained for singlet states is explained by the fact that the 6^1s state is only 40.5 cm^{-1} away from the 2-photon resonance.

Two general remarks have to be made. First, the multiphoton ionization cross section is very sensitive to the proximity of a resonance. For example, two of the values for σ_4 published in the literature for cesium differ by an order of magnitude at almost the same laser wavelength: $\sigma_4 = 7.5 \times 10^{-109}$ cm^8 s^3 at 1056 nm (Normand and Morellec, 1980) and $\sigma_4 = 1.0 \times 10^{-107}$ cm^8 s^3 at 1060 nm (Delone *et al.*, 1976). These values were obtained on either side of the resonance on the 6F state at 1059 nm. Second, some of the multiphoton ionization cross sections of alkaline atoms published in the literature may be somewhat misleading because of the contribution that dimers make to the atomic ion signal. This is due to the large ionization cross section of dimers relative to atoms. This problem will be considered in detail in Section IV and is especially important in the 2-photon ionization of alkaline atoms, and to a lesser extent in 3- and 4-photon ionization. From this point of view, the multiphoton ionization of rare gases at low density avoids the molecular problem completely.

Much experimental work has been undertaken on the nonresonant multiphoton ionization of rare gases. However, experimental ionization cross sections can rarely be checked against theory owing to the lack of relevant calculations. The N-photon ionization rate W, which varies with the laser intensity I, as I^N, characterizes much more precisely the nonlinear process than the generalized ionization cross section σ_N defined as being independent of I. The N-photon ionization of rare gases has been investigated in high

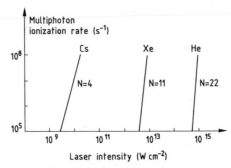

Fig. 3. Schematic diagram showing the laser intensity required to the 4-photon ionization of Cs, the 11-photon ionization of Xe, and the 22-photon ionization of He induced by coherent laser pulses at 1.06 μm.

laser intensity ranges, up to 10^{15} W cm^{-2}, corresponding to a laser electric field of the order of intra-atomic fields. In these ultrastrong laser fields, no departure from the I^N law was observed. For example, accurate measurements in 22-photon ionization of helium with a bandwidth-limited, 15-ps, laser pulse at 1.06 μm show a $I^{22 \pm 0.2}$ intensity dependence (Lompré et al., 1980b). Such an accuracy is only possible by measuring the spatial distribution of the focused laser intensity (Lompré et al., 1982). The intensity dependence of the multiphoton ionization rate has been found to be given by the lowest order term of the perturbation series, provided we are far from a resonance. This remark is still valid for a laser field of the order of the intra-atomic field, probably because laser-induced atomic level shifts keep moderate values. None of the experiments convincingly confirmed the tunneling ionization hypothesis (Keldysh, 1965), which was believed to dominate over multiphoton ionization in ultrastrong fields.

Emphasis should be given to the fact that N-photon ionization with large N values can be observed by simply increasing the laser intensity. As an example, Fig. 3 shows that, by using a coherent laser pulse at 1.06 μm, the 4-photon ionization of cesium occurs at 10^{10} W cm^{-2}, the 11-photon ionization of xenon at 10^{13} W cm^{-2}, and the 22-photon ionization of helium at 10^{15} W cm^{-2}, for the same ionization rate $W = 10^8$ s^{-1}.

B. Coherence Effects in Nonresonant Multiphoton Ionization

Multiphoton ionization of atoms is an inherently nonlinear process and as such depends not simply on the laser intensity but also on its coherence properties. Many multiphoton ionization experiments reported in the literature have been performed with incoherent laser pulses generated by multimode Q-switched lasers which had spectral bandwidths of about 1 cm^{-1} and strong temporal fluctuations. The duration of these peak intensities is given by $1/b$, where b is the spectral bandwidth of the laser pulse, i.e., 30 ps for $b = 1$ cm^{-1}. As was shown in the introduction, the characteristic ionization time of nonresonant multiphoton ionization of atoms can be as short as 10^{-15} s. As a result, atoms "see" the fluctuations in the "arrival" of photons and respond, not only to the average number of photons per unit time, but also to the way this number fluctuates. Photons arrive bunched in an incoherent laser pulse Fig. 4b, while they arrive in single file in a coherent laser pulse as shown in Fig. 4a. It is quite obvious that bunching effects change the time interval between photons and will directly affect the multiphoton absorption rate. This is referred to as the effect of correlations or photon statistics.

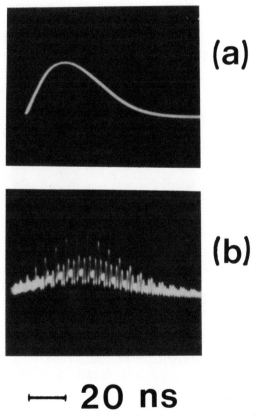

Fig. 4. Temporal distribution function of the laser intensity of a: (a) single-mode laser pulse; (b) multimode laser pulse with a 3-GHz bandwidth.

The study of the effects of intensity fluctuations or photon statistics began as early as 1964 in nonlinear optical processes (Ducuing and Bloembergen, 1964) and 1966 on 2-photon absorption (Lambropoulos *et al.*, 1966). Specific calculations on coherence effects in multiphoton ionization have been performed later (Debethune, 1972; Sanchez, 1975). The fundamental result of these calculations is that the rate of nonresonant N-photon ionization with chaotic light is larger by a factor $N!$ than with purely coherent light.

Let us describe the temporal fluctuations of a laser pulse used for multiphoton ionization experiments. The instantaneous laser intensity seen by atoms can be expressed in the form

$$I(t) = \bar{I}_M G(t) i(t) \tag{2.4}$$

where $\bar{I}_M$ is the maximum time-averaged intensity; $G(t)$ is the normalized temporal distribution function envelope of the laser intensity with a duration of about 10^{-8} s for a Q-switched laser pulse; and $i(t)$ is a periodic function, which will play a very fundamental role in this study. It has a stochastic pattern which depends on the number of modes and on both relative phases and amplitudes of modes. We generally measure only the time-averaged intensity

$$\bar{I}(t) = \bar{I}_M G(t) \tag{2.5}$$

without taking into account the peak intensity function $i(t)$. Whereas multiphoton ionization of atoms is a highly nonlinear process which is very sensitive to peak intensity function since the N-photon ionization rate is proportional to the Nth power of the instantaneous intensity, $i(t)$ will be called the peak intensity function.

Let us define the Nth-order time-independent peak intensity moment by

$$f_N = \langle \overline{i^N} \rangle \tag{2.6}$$

where the angle brackets stand for the ensemble average. The statistics of the laser pulse is readily characterized by f_N which can be easily related to multiphoton ionization. The number of ions induced by a multimode laser pulse is

$$N_m = \beta \int I^N(t)\, dt = \beta \int \bar{I}^N(t) i^N(t)\, dt \tag{2.7}$$

where β is a proportional factor.

The number of ions induced by a single-mode laser pulse having the same average intensity is

$$N_1 = \beta \int I^N(t)\, dt = \beta \int \bar{I}^N(t)\, dt \tag{2.8}$$

Hence in our experimental conditions,

$$\langle N_{\mathrm{m}}/N_1 \rangle = \langle \overline{I^N}/\overline{I}^N \rangle = f_N \tag{2.9}$$

Thus f_N may be related to multiphoton ionization very simply. It is the enhancement of the ion signal due to the multimode operation of the laser.

The theoretical predictions of an $N!$ enhancement in the nonresonant N-photon ionization rate have been experimentally corroborated. For example, Fig. 5 shows that the nonresonant 4-photon ionization rate of Cs atoms induced by an incoherent 3-GHz-bandwidth laser pulse is enhanced by 4! compared to the rate induced by a single-mode laser pulse of the same average intensity (Lompré *et al.*, 1981). Likewise, the nonresonant 5-photon ionization rate of Na atoms with an incoherent laser pulse is more efficient by a factor 5! than that encountered with a single-mode laser pulse of the same average intensity (Arslanbekov, 1976).

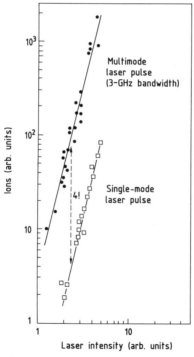

Fig. 5. Log–log plot of the variation of the number of ions formed as a function of laser intensity in the nonresonant 4-photon ionization of Cs. The ionization rate induced by an incoherent (3-GHz bandwidth) laser pulse is enhanced by 4! compared to that induced by a coherent laser pulse of the same average intensity.

The most dramatic experimental demonstration of these effects has been performed in the 11-photon ionization of xenon atoms by varying the mode-structure of a Nd-glass laser pulse from a single-mode to about one hundred modes (Lecompte *et al.*, 1975). The solid line in Fig. 6 shows the enhancement of the number of ions, i.e., the experimental value of the f_{11} moment when the number of modes is increased from 1 to 100 for the same average laser intensity. The number of ions is enhanced by nearly 10^7 when the number of modes is increased from 1 (coherence time 40 ns) to 100 (coherence time 8 ps). As a comparison, the dashed line in this figure also shows the f_{11} moment calculated in assuming that phases of the modes are independent. The difference between the experimental and the calculated f_{11} moment comes mainly from the fact that the laser spectrum consisted of several bands with ten modes in each band. Therefore, the statistical properties of this laser radiation were different from that used in calculating the f_{11} moment as-

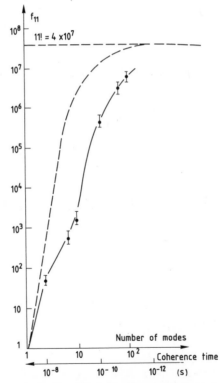

Fig. 6. The 11-photon ionization of Xe induced by varying the mode structure of a Nd-glass laser pulse. The solid line shows the enhancement of the number of ions, i.e., the experimental value of the f_{11} moment, when the number of modes is increased from 1 to 100. The dashed line shows the f_{11} moment calculated in assuming that phases of the modes are independent.

suming that phases of the modes are independent. As anticipated from Fig. 6, experimental and calculated f_{11} values tend towards each other and are expected to become identical at an asymptotic value 11! for a very large number of modes.

As a conclusion, the off-resonant N-photon ionization rate can be written as

$$w = \sigma_N f_N \bar{I}^N \tag{2.10}$$

The Nth-order peak intensity moment, or Nth-order autocorrelation function f_N, is equal to unity for a single-mode laser pulse, or a bandwidth-limited pulse, i.e., a pulse completely devoid of intensity modulation. In the limit of an infinite number of independent modes, f_N equals $N!$. The autocorrelation function f_N depends both on the laser spectral bandwidth and on the order N of the nonlinear process. It is a small correction factor (≤ 2) for a 2-photon process, while it has a dramatic effect in high-order nonlinear processes such as multiphoton ionization of rare gases. This effect can give, in many cases, the key to the explanation of discrepancies between σ_N calculated in assuming coherent laser radiation and σ_N measured with incoherent laser pulses. Conversely, multiphoton ionization processes allow us to consider an atom irradiated by a laser pulse, as an ideal photon detector concerning the statistical properties of the laser pulse. The determination of the Nth-order autocorrelation function of the laser peak intensity is of special interest to fully characterize a laser radiation.

III. RESONANCE EFFECTS IN MULTIPHOTON IONIZATION OF ATOMS

By tuning the laser frequency, the multiphoton ionization rate of atoms can be made to exhibit a typical resonant character when the energy of an integral number of photons approaches the energy of an allowed atomic transition. W increases dramatically while retaining a finite value because of damping terms arising from the coupling of the resonant state with the ground state and the continuum.

A. Resonance Effects in a Moderate Laser Intensity Range (10^7–10^9 W cm^{-2})

1. Resonance Effects Induced by Coherent Laser Pulses

The characteristics of the resonance effects have been extensively investigated in the 4-photon ionization of Cs atoms with a Nd-glass laser pulse and the tuning of the frequency through the resonant 3-photon transition

6S → 6F (Grinchuk *et al.*, 1975; Morellec *et al.*, 1976; Petite *et al.*, 1979; Lompré *et al.*, 1978). This example has been chosen for two reasons. First, calculations can be carried out with a high degree of accuracy for Cs atoms (Gontier and Trahin, 1979a, Georges and Lambropoulos, 1977; Crance and Aymar, 1979). Second, the Nd-glass laser is very well suited to perform this experiment at 1059 nm because it can deliver a coherent pulse and the wavelength can be tuned within the 1052–1065-nm range. It is of interest to use a coherent pulse because resonance effects can be investigated without any mixing with laser coherence effects and laser bandwidth effects so that a direct comparison with theory can easily be made. Furthermore, it should be pointed out that coherent pulses from high-power dye laser were not available a few years ago.

As we have seen previously, the ionization time of a quasi-resonant multiphoton ionization process is given by $1/\Delta E$, i.e., 10^{-11} s for $\Delta E = 3$ cm^{-1}. Here we are concerned with resonance processes at a laser intensity of 10^8 W cm^{-2}; the ionization time is much longer and is given by the lifetime of the resonant state induced by the laser field. With our experimental conditions, the 6F resonant state is much more strongly coupled to the continuum than to the ground state. This means that the photoionization rate from the 6F level is much greater than the decay rate to the ground state due to stimulated emission. At the exact resonance, the ionization time

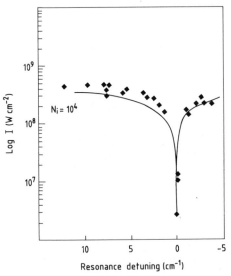

Fig. 7. Laser intensity I required to form 10^4 atomic Cs ions as a function of resonance detuning of the 3-photon transition 6S → 6F in 4-photon ionization. Experimental points are from Morellec *et al.* (1976) and the solid line is a calculation from Crance (1978).

$T = 1/(\sigma I)$, where $\sigma = 1.4 \times 10^{-8}$ cm^2 is the photoionization cross section of the 6F level. Ionization time $T \simeq 10^{-9}$ s for $I = 10^8$ W cm^{-2}. Thus the resonant multiphoton ionization process requires a laser intensity, roughly 100 times lower than for the quasi-resonant process, in agreement with the ratio of the aforementioned times $(10^{-9}/10^{-11})$. This point is verified by the results shown in Fig. 7, which shows the laser intensity I required to form 10^4 ions as a function of resonance detuning. The full curve is derived from a theoretical calculation (Crance, 1978) and the experimental points have been obtained using a single-mode laser pulse (Morellec et al., 1976).

Conversely, if the laser intensity is kept fixed, resonance effects will be expressed as an enhancement in the number of ions at the resonance frequency, as shown in Fig. 8. This result was obtained using a bandwidth-limited 15-ps laser pulse (Lompré et al., 1978). It clearly demonstrates a shift in the resonance profiles when the laser intensity is increased from 10^8 to 10^9 W cm^{-2}. As is well known, the exchange of photons between laser radiation and atoms shifts and broadens atomic levels. These effects have been well described within the framework of the dressed atom theory (Cohen-Tannoudji, 1967). Atomic level shifts induce a resonance shift as

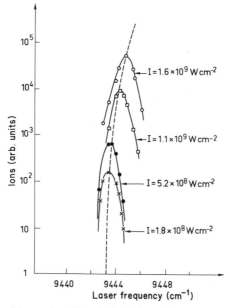

Fig. 8. Variation of the number of ions in the 4-photon ionization of Cs as a function of laser frequency in the neighborhood of the resonant 3-photon transition 6S → 6F. The dashed line shows the resonance shift for increasing values of laser intensity I.

shown in Fig. 8. This resonance shift is linear with respect to the laser intensity I, $\Delta = \alpha I$ with $\alpha = 2$ cm^{-1}/GW cm^{-2}, in excellent agreement with calculations (Crance, 1978; Gontier and Trahin, 1978) as shown in Fig. 4 in the theoretical chapter written by Gontier and Trahin (Chapter 3, this volume).

The importance of atomic level shifts induced by the laser field is also emphasized through the law describing the variation in the number of ions N_i as a function of the laser intensity I very near to the resonance. Let us consider a laser frequency that gives rise to a small static resonance detuning $\Delta E = E_{6F} - E_{6S} - 3E_p$, where E_p is the photon energy. Atomic level shifts bring about a dynamic resonance detuning $\delta E = \Delta E + \alpha I$, which can tune the resonance to be closer or farther away, depending upon whether ΔE is positive or negative. Consequently, the variation in the number of ions formed will be faster or slower than the simple I^4 law. We observed indeed an I^K law with K different from N. Figure 9 shows the variation of $K = (\partial \log N_i)/(\partial \log I)$ as a function of ΔE. This result has been obtained using a single-mode Nd-glass laser pulse (Morellec et al., 1976). In this representation, a resonance appears as a sharp phenomenon marked by a dramatic change in the nonlinear order K, which no longer corresponds to the number of photons absorbed by the atom. For ΔE values larger than 10 cm^{-1}, atomic level shifts become insignificant compared to ΔE; the I^4 law is again valid and characterizes the offresonance 4-photon ionization process. It should be pointed out that a very good agreement between theory and experiment is obtained as shown in Fig. 6 in Chapter 3 by Gontier and Trahin.

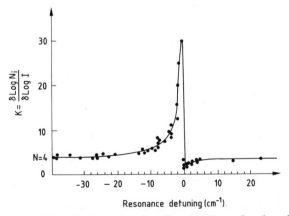

Fig. 9. Variation of the effective order of nonlinearity K as a function of the resonance detuning ΔE of the resonant 3-photon transition 6S → 6F in the 4-photon ionization of Cs.

As far as the width (FWHM) of the resonance profile is concerned, Fig. 8 shows a width of 1 cm^{-1} governed by the broad laser bandwidth (1.4 cm^{-1}) of the 1.5×10^{-11} s pulse. The situation is different when a single-mode laser pulse with a duration of 3×10^{-8} s and a narrow bandwidth of 20 MHz is used. Figure 10 obtained with such a laser pulse shows four resonance peaks due to a 0.3 cm^{-1} hyperfine structure of the 6S ground state of the Cs atom and to a 0.1 cm^{-1} fine structure of the 6F resonant state (Petite *et al.*, 1979). The width of each of the four resonance peaks is 600 MHz, much broader than the 20 MHz laser bandwidth. This is due to both an intensity independent Doppler broadening, which contributes 300 MHz, and an intensity dependent broadening because the laser intensity is inhomogeneous in the ionization region. There is a distribution of intensity between zero and a maximum value. As the resonance shift is linear with respect to the laser intensity, we also have a large distribution in the shifts, which appears as a broadening in the resonance profiles and contributes 300 MHz for $I = 2 \times 10^7$ W cm^{-2}. This intensity dependent broadening increases when the laser intensity is increased, as shown in Fig. 10. That is in good agreement with a calculation performed by Gontier and Trahin (1979) and shown in Fig. 3 in their theoretical chapter (Chapter 3).

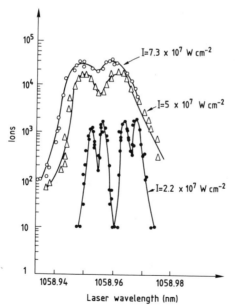

Fig. 10. Resonance profiles in the 4-photon ionization of Cs for three values of laser intensity *I*. Hyperfine structure of the ground state and the fine structure of the 6F resonant state are well resolved for the lowest intensity value.

2. Temporal Effects in Resonant Multiphoton
Ionization Processes

Temporal aspects induced by the laser pulse duration play an important role in resonance effects. Figure 11a shows schematically the time evolution of the population of the resonant state n_r (Lompré et al., 1980b). Figure 11b shows the time evolution of the corresponding multiphoton ionization probability P. Both figures exhibit three successive temporal regions. The population of the resonant state n_r increases as t^2 in region I, reaches a stationary regime in region II, and decays exponentially to zero in region III. As far as the ionization probability P is concerned, the temporal evolution obeys a t^3 law in the first region and a linear law in the second region before reaching an ionic saturation (due to the total ionization of all the atoms) in region III. It should be pointed out that a time-independent ionization rate, i.e., an ionization probability per unit time, can be defined only in the second region. The population of the resonant state n_r is given by the relation

$$n_r = \frac{|R_{gr}|^2}{(\sigma I)^2 + (\Delta E)^2} \qquad (3.1)$$

where σ is the photoionization cross section of the resonant state and $|R_{gr}|^2$ is the matrix element of the transition between the ground state and the

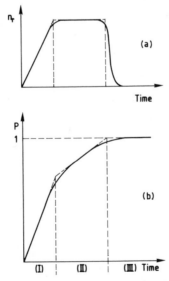

Fig. 11. Time evolution of (a) the population n_r of the resonant state and (b) the multiphoton ionization probability P in the resonant multiphoton ionization of an atom. ΔE is the resonance detuning and σ is the photoionization cross section of the resonant state.

resonant state. The multiphoton ionization probability is proportional to

$$\frac{|R_{\mathrm{gr}}|^2 \sigma I t}{(\Delta E)^2 + (\sigma I)^2} \tag{3.2}$$

Both relations are valid for $\sigma I \gg |R_{\mathrm{gr}}|$.

3. Resonance Effects Induced by Incoherent Laser Pulses

The characteristic ionization time in nonresonant multiphoton ionization of atoms can be as short as 10^{-15} s. As a result, atoms "see" the fluctuations in the laser intensity and are very sensitive to the statistical properties of laser radiation. In resonant multiphoton ionization of alkaline atoms in moderate laser intensity range (10^7–10^9 W cm^{-2}), the characteristic ionization time can be as long as 10^{-9} s. Consequently, the statistical properties of laser radiation are not expected to enhance dramatically the resonant multiphoton ionization rate. However, the resonance shift due to laser-induced atomic level shifts is linear as a function of the laser intensity and can be significantly enhanced by laser intensity fluctuations of an incoherent laser pulse. In addition, the laser bandwidth begins to play a role as soon as it becomes comparable to the resonance detuning. Therefore, laser-temporal coherence effects cannot be investigated independently of laser bandwidth effects in resonant multiphoton ionization.

In the past few years, a number of authors have investigated this problem theoretically (Kovarsky *et al.*, 1976; Armstrong *et al.*, 1976; Agostini *et al.*, 1978). It is only very recently that specific calculations have been performed (Gontier and Trahin, 1979b, Dixit and Lambropoulos, 1980; Zoller and Lambropoulos, 1980). Recently, laser coherence effects and bandwidth effects have been investigated experimentally in 4-photon ionization of Cs, with a 3-photon resonance on the 6F level (Lompré *et al.*, 1981). The sophisticated laser used in this experiment allows both wavelength and temporal coherence to be changed by varying the number of longitudinal modes. Increasing the number of modes leads to stronger intensity fluctuations and a larger bandwidth. The resonance curves obtained with incoherent laser pulses are observed to be shifted and broadened with regard to those induced by coherent pulses with the same average intensity. The statistical enhancement of the resonance shift is shown directly in Fig. 12. This figure gives two resonance profiles obtained at the same laser intensity $\bar{I} = 5.6 \times 10^7$ W cm^{-2} with a coherent pulse (curve A) and an incoherent pulse (curve B), which has a bandwidth $b = 0.1$ cm^{-1}. In the resonance profile obtained with the multimode laser pulse, the 6F fine structure is not resolved because the laser bandwidth is larger. The important result of this figure is the additional shift induced by the incoherent laser pulse. This additional shift is 0.2 cm^{-1}

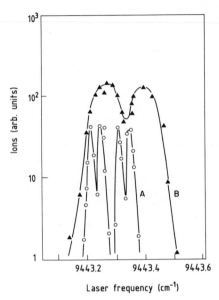

Fig. 12. Resonance profiles obtained in the 4-photon ionization of Cs atoms with a single-mode laser pulse (curve A) and a multimode laser pulse with a bandwidth $b = 3$ GHz (curve B). The laser intensity for both profiles is 5.6×10^7 W cm^{-2}.

expressed in terms of the energy of the 3-photon transition 6S → 6F. The statistical enhancement of the resonance shift is clearly demonstrated in Fig. 13, which shows the two laws of variation of the resonance shift as a function of the laser intensity with a coherent pulse (dashed line) and an incoherent laser pulse (solid line). The difference between the two lines gives the additional shift due to the laser intensity fluctuations, i.e., (3.6 ± 0.3) cm^{-1} for $I = 1$ GW cm^{-2}. This experimental result is in excellent agreement with a previous calculation (Gontier and Trahin, 1979b). Figure 13 shows that the resonance shift induced with an incoherent laser pulse is enhanced by a factor 2.8 ± 0.2 compared with the shift induced with a coherent pulse. This result is very well explained by zero bandwidth theoretical models. However, large-bandwidth models do not describe experimental results so well.

The width of the resonance profiles induced by multimode laser pulses depends on the laser bandwidth, the intensity-dependent broadening, and the statistical broadening. Figure 14 shows the variation of the resonance widths as a function of the laser intensity for single-mode laser pulses (curve A) and multimode laser pulses with bandwidths 3×10^{-2} cm^{-1} (curve B), 8×10^{-2} cm^{-1} (curve C), and 0.15 cm^{-1} (curve D). At low intensities the resonance width is independent of intensity and is governed by the laser bandwidth while at high intensities it depends on statistical broadening. These results are in qualitative agreement with different calculations.

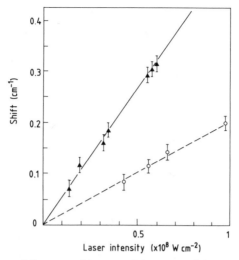

Fig. 13. Resonance shift, expressed in terms of the energy of the 3-photon transition 6S →
6F, as a function of laser intensity for a single-mode laser pulse (dashed line) and an incoherent
laser pulse (solid line).

B. Resonance Effects in a High Laser Intensity
   $(10^{13} \text{ W cm}^{-2})$

Resonance effects in the 10^{13} W cm^{-2} intensity range can be investigated
in the multiphoton ionization of rare gases using a tunable-wavelength
coherent Nd-glass laser pulse. We consider the 12-photon ionization of

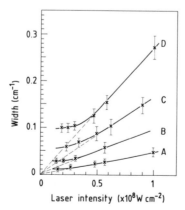

Fig. 14. Variation of the resonance width as a function of laser intensity for single-mode laser
pulses (curve A) and multimode laser pulses with bandwidths 3×10^{-2} cm^{-1} (curve B), 8 ×
10^{-2} cm^{-1} (curve C), and 0.15 cm^{-1} (curve D).

krypton when a resonance occurs on a 5d or 4d' state with the penultimate photon absorbed. When the laser intensity used is high enough, no significant enhancement in the number of ions N_i is observed, although the resonance effects exhibit a large variation in the effective order or nonlinearity $K = (\partial \log N_i)/(\partial \log I)$ very near to the resonance (Lompré et al., 1980b).

In contrast with the very good agreement between theoretical and experimental results on resonance effects in the 4-photon ionization of Cs in the intensity range of 10^7–10^9 W cm^{-2}, the experimental results on resonance effects for the 12-photon ionization of krypton at 10^{13} W cm^{-2} cannot, as yet, be quantitatively explained because of the lack of relevant calculations. The absence of such calculations is mainly owing to the lack of atomic data and especially oscillator strengths for rare gases. However, we can give a qualitative explanation in terms of the variation of multiphoton ionization probability in the close vicinity of the resonance, as was shown in Section III.A.2. The photoionization cross section of the resonant state σ is not known with any accuracy for 5d or 4d' states in krypton. It is estimated to be between 10^{-20} and 10^{-19} cm^2. In Fig. 15 the resonant multiphoton ionization probability is plotted for 10^{12} and 10^{13} W cm^{-2}, assuming a photoionization cross section $\sigma = 10^{-20}$ cm^2. For $I = 10^{13}$ W cm^{-2}, no significant enhancement is observed, which explains why no significant resonance profile is experimentally observed under the same conditions. This flattening

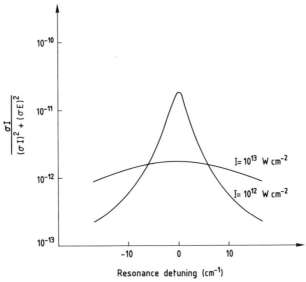

Fig. 15. Schematic representation of a resonance profile at laser intensity 10^{12} and 10^{13} W cm^{-2}.

of the resonance profile arises from the dominant contribution of the $(\sigma I)^2$ term due to the high I value over the $(\Delta E)^2$ term throughout a very broad resonance detuning. Conversely, if we consider examples with a significantly weaker value of laser intensity, or a resonant state in which the photoionization cross section is smaller, a resonance profile with a small amplitude might still be observed, as in a previous experiment on the 11-photon ionization of Xe performed at the Levedev Institute by Alimov and Delone (1976). In this experiment, a resonance profile was observed with an enhancement of ten at the most. The measurement of such a resonance profile, together with a knowledge of the laser intensity as an absolute value, would allow us to accurately determine the photoionization cross section for a given atomic state. Such data are often missing in the literature.

The dramatic change in the effective order of nonlinearity $K = (\partial \log N_i)/(\partial \log I)$ is also observed very near to a resonance in krypton at 10^{13} W cm^{-2}, even when the resonance profile is too flat to be measured. The variation of K is explained, as for Cs at 10^8 W cm^{-2}, in terms of atomic level shifts induced by the laser field. However, at 10^{13} W cm^{-2} calculations would have to take into account not only the linear term αI but also higher order terms βI^2, γI^3, ..., which are expected to limit shift values.

IV. ANTIRESONANCE EFFECTS IN TWO-PHOTON IONIZATION OF Cs ATOMS

As is well known, the 2-photon ionization cross section σ_2 at laser frequency ω is given by second-order perturbation theory

$$\sigma_2 \propto \left| \sum_n \frac{\langle f|r|n\rangle \langle n|r|g\rangle}{\omega_n - \omega} \right|^2 \tag{4.1}$$

and it is necessary to sum all the intermediate states $|n\rangle$ of energy $\hbar\omega_n$. Figure 16 shows the variation of the 2-photon ionization cross section for Cs atoms, calculated by Bebb (1966), as a function of the photon energy. The variation of σ_2 exhibits both successive resonance profiles arising from minima in the denominator of the expression for σ_2 and deep minima due to destructive interference among the terms of the sum. Second-order perturbation theory predicts a first minimum for a 2.6-eV photon energy, i.e., about 480 nm, arising from cancellation of the 6P and the other nP states with a main contribution from the 7P states.

It was expected that this minimum in the 2-photon ionization cross section for Cs atom would be difficult to investigate experimentally because of the contribution dimers make to the Cs$^+$ signal through processes such as

$$Cs_2 + 2hv \rightarrow Cs_2^+ + e \tag{4.2}$$

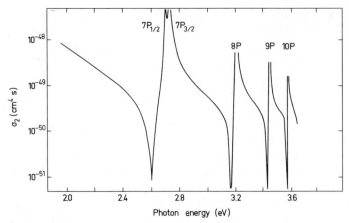

Fig. 16. Variation of the 2-photon ionization cross section for Cs atoms as a function of the photon energy, calculated by Bebb (1966).

followed by

$$Cs_2^+ + h\nu \rightarrow Cs^+ + Cs \tag{4.3}$$

Because the density of resonant states is high, such processes can be enhanced resonantly, especially at laser wavelengths in the 480-nm region where there is a strong Cs_2 absorption band:

$$C^1\Pi_u \leftarrow X^1\Sigma_g^+$$

Atomic Cs^+ ions resulting from these molecular processes are indistinguishable from Cs^+ ions originating from the 2-photon ionization of Cs atoms. The molecular density n_{Cs_2} is much lower than the atomic density n_{Cs}, typically $n_{Cs_2}/n_{Cs} = 5 \times 10^{-4}$. However the 2-photon ionization and dissociation rates for Cs_2 are much higher than the 2-photon ionization rate for Cs atoms. Consequently, it is necessary to minimize the dimer density. This can be partially achieved through a thermal dissociation in a superheater. The molecular component reduction factor is roughly 20 (Morellec *et al.*, 1980). There is no possibility whatsoever of dissociating all the dimers by this process.

The most important parameter which makes this experiment feasible is the laser intensity. It has been clearly demonstrated in a previous experiment on 4-photon ionization of Cs atoms at 1.06 μm that the contribution dimers make to the Cs^+ signal plays a dominant role with laser intensities less than 10^7 W cm^{-2} (Held *et al.*, 1972). It is therefore necessary to use a high intensity in the 10^9-10^{10} W cm^{-2} range. The problem can be solved by taking advantage of the different intensity dependencies of the number of atomic Cs ions produced by the 2-photon ionization of Cs atoms that obey an I^2 law, and the number of atomic Cs ions originating from the different molecular channels that become saturated at high laser intensity and obey an I^K law, with $K \lesssim 1$. The contribution dimers make to the Cs$^+$ signal therefore dominates completely in the low laser intensity range, while the Cs$^+$ signal originating from the 2-photon ionization of Cs atoms dominates the dimer contribution when the laser intensity is higher than 10^9 W cm^{-2}.

The 2-photon ionization cross section of atomic cesium was measured in the 460–540 nm range in an experiment that fulfills the following two requirements: a laser intensity of 10^{10} W cm^{-2} and a partial thermal dissociation of dimers in a superheater (Morellec $et\ al.$, 1980). Figure 17 shows that σ_2 exhibits a deep minimum, as predicted by second-order perturbation theory. This result has two important consequences.

First, it confirms the validity of second-order perturbation theory and clarifies a puzzling situation which arose from a previous experiment. In this experiment there was no minimum in the 2-photon ionization cross section of cesium and ionization cross section values occured that were of up to four orders of magnitude higher than theoretical values (Granneman and Van der Wiel, 1975; Klewer $et\ al.$, 1977). This result can be explained because the experiment was performed in a very low intensity range (10^4–10^5 W cm^{-2})

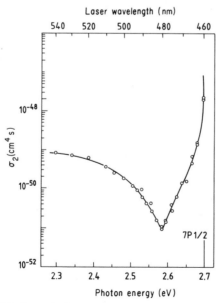

Fig. 17. Measured 2-photon ionization cross section of Cs atoms as a function of photon energy and laser wavelength.

where the contribution of dimers to the Cs^+ signal dominates completely the Cs^+ signal originating from the 2-photon ionization of Cs atoms. However, this result brought about a confused situation (Knight, 1977) and even led some theoretical physicists to question the validity of perturbation theory. Finally, the observation of the minimum in the 2-photon ionization cross section of Cs atoms, shown in Fig. 17, as well as in the 3-photon ionization of K atoms (Delone et al., 1981), makes the situation clear and confirms once and for all the validity of perturbation theory in this intensity range.

Second, the accuracy of the results shown in Fig. 17 allows very profitable comparisons to be made with several calculations using second-order perturbation theory. It should be pointed out that in the close vicinity of a resonance with an intermediate state, the 2-photon ionization cross section is determined from only a few matrix elements. For example, in the resonant 2-photon ionization of Cs atoms on the 7P state, it has been shown that the single intermediate 7P state approximation is valid within a factor of 2 for several hundred wave numbers on either side of the resonance (Lambropoulos and Teague, 1976). At or very near to a minimum, the 2-photon ionization cross section is determined from a large number of matrix elements because a large number of terms manifest cancellation, producing the minimum. The exact position and depth of the minimum can only be determined by an exact infinite summation of terms with exact matrix elements. All the

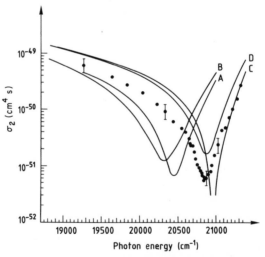

Fig. 18. 2-photon ionization cross section of Cs atoms as a function of photon energy. Comparison of experimental results (Morellec et al., 1980) with different calculations using the quantum defect method (curves A and B) and the model potential theory (curves C and D).

calculations published in the literature predict a minimum in the 2-photon ionization cross section of Cs atoms, but they disagree on the position and the depth of the minimum. Therefore, the experimental result can be a very sensitive test of the accuracy of the different calculation models. The comparison of the experimental results with calculations (Teague *et al.*, 1976; Lambropoulos and Teague, 1976; Declemy *et al.*, 1981; Aymar and Crance, 1982) is shown in Fig. 18. This figure demonstrates that the model potential theory [curves C (Teague *et al.*, 1976) and D (Aymar and Crance, 1982)] gives much more accurate results than the quantum defect theory [curves A (Lambropoulos and Teague, 1976) and B (Declemy *et al.*, 1981)]. This is because the first P levels give the main contribution to σ_2 at the minimum, while the quantum defect method gives matrix elements of poor accuracy for the first P levels. Furthermore, a still better agreement between experimental result and the model potential theory is obtained by taking into account the continuum contribution (Aymar and Crance, 1982).

V. CONCLUSION AND FUTURE PROSPECTS

The different aspects of the multiphoton ionization of atoms are now well understood and many of them can be correctly described by rigorous theoretical models. For example, accurate measurements of absolute values of the 2-, 3-, and 4-photon ionization cross sections of alkaline atoms and metastable helium atoms are in good agreement with calculated values, and clearly emphasize the validity of perturbation theory. In the same way, there is also a good agreement between theoretical and experimental results on resonance effects in multiphoton ionization of cesium atoms in the moderate laser intensity range, $10^7 - 10^9$ W cm^{-2}. Resonance effects emphasize the important role played by laser-induced atomic level shifts. Destructive interference effects, which give rise to minima in the multiphoton ionization cross sections of atoms, have been investigated with a high degree of accuracy in the 2-photon ionization of Cs atoms. This effect has been successfully used to check the validity of different calculations models. Lastly, laser temporal-coherence effects increase the nonresonant N-photon ionization rate by $N!$, in good agreement with theoretical calculations. Combined coherence and resonance effects yield a significant enhancement in both the laser-induced resonance shift and resonance width.

It should be pointed out that the relative influences of resonance effects and coherence effects vary with the number N of photons involved in the ionization process and consequently with the required laser intensity.

In the moderate laser intensity range corresponding to 2- or 3-photon ionization of atoms, resonance effects play a dominant role, while photon statistics of the laser radiation are relatively unimportant (they contribute a factor of 2 at the most in a nonresonant 2-photon process). On the contrary, in the very high intensity range required to observe large Nth-order ionization processes, coherence effects play a dramatic role through the $N!$ effect, while resonance effects are so highly damped as a result of the high laser intensity that resonance profiles are no longer observed.

Now that we have a synthetic understanding of multiphoton absorption processes in atoms, it would be of interest to extend it to new fields where multiphoton processes can play a role. Different topics can be considered.

(1) Molecular spectroscopy can now benefit considerably from multiphoton ionization processes. New states can be reached and identified, especially through the optical double resonance multiphoton ionization spectroscopy of molecules. This technique allows a significant simplification of the spectrum produced by allowing a tunable-wavelength laser to bring the molecule to a resonant intermediate state through a 2- or 3-photon excitation process, with a second laser to provide the ionization step.

Furthermore, multiphoton processes in molecules leading to excitation, ionization, or dissociation have been the subject of a great deal of experimental and theoretical work but are far from being satisfactorily understood. It is still quite an open field.

(2) Competition between resonant multiphoton ionization of atoms or molecules and ultraviolet odd-harmonic generation at high pressure is a new promising field. Recent experiments have shown that resonant multiphoton ionization of atoms or molecules vanishes to the benefit of a vacuum ultraviolet third-harmonic generation when the atomic or molecular pressure is above approximately 1 Torr (Miller *et al.*, 1980). This topic is of interest for two reasons. First, from a basic point of view, it can lead to an understanding of the cooperative effects involved, together with the possible optical bistability occurring in these collective atomic or molecular systems. On the other hand, such effects could lead to the production of promising sources of coherent vacuum ultraviolet light for the vacuum ultraviolet spectroscopy of molecules.

(3) The production of multiply charged ions formed by multiphoton absorption in rare-gas atoms is a recently innovated topic. Up to quadruply charged ions can be observed with krypton atoms irradiated with a very high laser intensity (L'Hullier *et al.*, 1982). Multiply charged ions appear to be formed through multiphoton absorption from the atom's ground state. This leads to the challenge of calculating multiphoton ionization rates for many-electron atoms. This quite new topic could lead to interesting new developments. For example, advantage could be taken of the well-known radiative properties of these ions to generate extreme ultraviolet radiation.

Finally, other topics involving multiphoton absorption processes are considered separately in other chapters of this book.

(1) Multiphoton free–free transitions detected in the scattering of electrons on atoms or molecules, in the presence of a strong CO_2 laser field.

(2) Multiphoton absorption processes above the ionization threshold of an atom or a molecule.

(3) Creation of doubly charged ions in the multiphoton ionization of alkaline-earth atoms. Possible resonances on two-electron bound states or an autoionizing states could explain the creation of doubly charged ions.

REFERENCES

Agostini, P., Georges, A., Wheatley, S., Lambropoulos, P., and Levenson, M. (1978). *J. Phys. B* **11**, 1733.
Alimov, D., and Delone, N. (1976). *Sov. Phys. JETP (Engl. Transl.)* **43**, 15.
Armstrong, L., Lambropoulos, P., and Rahman, N. (1976). *Phys. Rev. Lett.* **36**, 952.
Arslanbekov, T. (1976). *Sov. J. Quantum Electr. (Engl. Transl.)* **6**, 117.
Aymar, M., and Crance, M. (1982). *J. Phys. B* **15**, 719.
Bebb, H. (1966). *Phys. Rev.* **149**, 25.
Cohen-Tannoudji, C. (1967). *Cargese Lect. Phys.* **2**, 347.
Crance, M. (1978). *J. Phys. B* **11**, 1931.
Crance, M., and Aymar, M. (1979). *J. Phys. B* **12**, 3665.
Crance, M., and Aymar, M. (1980). *J. Phys. B* **13**, 4129.
Debethune, J. L. (1972). *Nuovo Cimento Soc. Ital. Fis. B* **12**, 101.
Declemy, A., Rachman, A., Jaouen, M., and Laplanche, G. (1981). *Phys. Rev. A* **23**, 1823.
Delone, G., Manakov, N., Preobrazhenskii, M., and Rapoport, L. (1976). *Sov. Phys. JETP (Engl. Transl.)* **43**, 642.
Delone, N., Alimov, D., Khabibulayen, P., Tursunov, M., and Preobrazhensky, M. (1981). *Opt. Commun.* **36**, 459.
Dixit, S., and Lambropoulos, P. (1980). *Phys. Rev. A* **21**, 168.
Ducuing, J., and Bloembergen, N. (1964). *Phys. Rev.* **133**, 1493.
Georges, A., and Lambropoulos, P. (1977). *Phys. Rev. A* **15**, 727.
Gold, A. (1969). "Quantum Optics," (R. J.. Glauber, ed.), p. 397, Italian Physical Society Course 42. Academic, New York.
Gontier, Y., and Trahin, M. (1978). *J. Phys. B* **11**, L131.
Gontier, Y., and Trahin, M. (1979a). *Phys. Rev. A* **19**, 264.
Gontier, Y., and Trahin, M. (1979b). *J. Phys. B* **12**, 2123.
Gontier, Y., and Trahin, M. (1980). *J. Phys. B* **13**, 259.
Granneman, E., and Van der Wiel, M. (1975). *J. Phys. B* **8**, 1617.
Grinchuk, V., Delone, G., and Petrosyan, K. (1975). *Sov. J. Plasma Phys. (Engl. Transl.)* **1**, 172.
Held, B., Mainfray, G., Manus, C., and Morellec, J. (1972).. *Phys. Rev. Lett.* **28**, 130.
Keldysh, L. (1965). *Sov. Phys. JETP (Engl. Transl.)* **20**, 1307.
Klewer, M., Beerlage, M., Granneman, E., and Van der Wiel, M. (1977). *J. Phys. B* **10**, L243.
Knight, P. (1977). *Nature (London)* **270**, 561.
Kovarsky, V., Perelman, N., and Todirashku, S. (1976). *Sov. J. Quantum Electron. (Engl. Transl.)* **6**, 980.
Lambropoulos, P., and Teague, M. (1976). *J. Phys. B* **9**, 587.

Lambropoulos, P., Kikuchi, C., and Osborn, R. (1966). *Phys. Rev.* **144**, 1081.

Lecompte, C., Mainfray, G., Manus, C. and Sanchez, F. (1975). *Phys. Rev. A* **11**, 1009.

L'Hullier, A., Lompré, L-A., Mainfray, G., and Manus, C. (1982). *Phys. Rev. Lett.* **48**, 1814.

Lompré, L. A., Mainfray, G., Manus, C., and Thebault, J. (1978). *J. Phys. (Orsay, Fr.)* **39**, 610.

Lompré, L. A., Mainfray, G., Mathieu, B., Watel, G., Aymar, M., and Crance, M. (1980a). *J. Phys. B* **13**, 1799.

Lompré L. A., Mainfray, G., and Manus, C. (1980b). *J. Phys. B* **13**, 85.

Lompré L. A., Mainfray, G., Manus, and Marinier, J. P. (1981). *J. Phys. B* **14**, 4307.

Lompré L. A., Mainfray, G., and Thebault, J. (1982). *Rev. Phys. Appl.* **17**, 21.

Miller J., Compton, R., Payne, M., and Garrett, W. (1980). *Phys. Rev. Lett.* **45**, 114.

Morellec, J., Normand, D., and Petite, G. (1976). *Phys. Rev. A* **14**, 300.

Morellec, J., Normand, D., Mainfray, G., and Manus, C. (1980). *Phys. Rev. Lett.* **44**, 1394.

Morellec, J., Normand, D., and Petite, G. (1982). *Adv. At. Mol. Phys.* **18**, 97.

Normand, D., and Morellec, J. (1980). *J. Phys. B* **13**, 1551.

Petite, G., Morellec, J., and Normand, D. (1979). *J. Phys. (Orsay, Fr.)* **40**, 115.

Rachman, A., Declemy, A., Jaouen, M., and Laplanche, G. (1979). *Europhys. Study Conf. Multiphoton Processes. Benodet, France, June 18–22, 1979.*

Sanchez, F. (1975). *Nuovo Cimento Soc. Ital. Fis B* **27**, 305.

Teague, M., Lambropoulos, P., Goodmanson, D., and Norcross, D. (1976). *Phys. Rev. A* **14**, 1057.

Zoller, P., and Lambropoulos, P. (1980). *J. Phys. B* **13**, 69.

3

Theory of Multiphoton Ionization of Atoms

Y. GONTIER AND M. TRAHIN

Service de Physique des Atomes et des Surfaces
Centre d'Etudes Nucléaires de Saclay
Gif-sur-Yvette, France

I. INTRODUCTION

In this chapter we present a theoretical discussion of nonresonant and resonant multiphoton ionization of atoms, which are two aspects of the same basic process in which bound electrons absorb several photons from the radiation field and go beyond the ionization threshold.

A convenient approach to multiphoton ionization (MPI) is to summarize all the information about the interaction between the field and the atom within an operator which allows us to determine the state of the system at any time.

In general, the time evolution operator $U(t)$ is not known. To calculate it we resort to perturbative techniques which give $U(t)$ in the form of a power series in the interaction V. This can be done on $U(t)$ itself by which we obtain a series where all the terms are time dependent. An alternative is to calculate, within the (time-independent) Schrödinger picture, an operator from which one determines perturbatively (or not) the operator $U(t)$. This method, which is suitable if one wishes to make a re-summation of the perturbation series, will be extensively used throughout this chapter.

When the series giving $U(t)$ converges, i.e., if processes involving more and more photon absorption and emission become less and less probable, we only consider the lowest-order nonvanishing term of the series. For MPI this term just contains the number of photon absorptions required by the ionization of the atom. The resulting probability is linear in time and enables us to define a time-independent transition rate (single-rate approximation).

Even within this lowest-order perturbation theory (LOPT), the calculation of high-order transition matrix elements requires the evaluation of infinite sums running over all the (real) atomic states. Such a difficulty has been circumvented in the literature by using different methods. Some methods are approximate, others provide exact results in atoms for which the potential is analytically or numerically known. The latter techniques are to be used when no particular state gives a prevalent contribution to the probability.

Note that when the frequency of the field is a submultiple of the energy difference between the ground state and any excited state, the probability becomes infinite and we must deal with a resonance. In LOPT, these undamped resonances give rise to a divergence of the perturbation series because many higher-order channels leading to the same final state can give larger contributions than that which is considered. It is necessary to take into account such contributions.

In general, the first few higher-order terms do not provide a good description of the resonance mechanism, principally in a region beyond the radius of convergence of the series. We must sum the whole perturbation series and find the expression of this sum which can be considered as the analytic continuation of the series outside its domain of convergence. Some theories involving renormalized Green's functions have taken into account a certain class of higher-order terms. Unfortunately, they contain multiple diagram counting and are irrelevant. In Section II we give the formulas resulting from a summation of the perturbation series where all the diagrams are counted once.

In Section III, we describe one of these exact summation techniques where the matrix elements are worked out in the coordinate representation.

In Section IV the dependence of the probability with regard to the photon frequency, the field intensity, and the interaction time is discussed within this framework. The behavior of the shift, the width, and the amplitude predicted by the theory is in excellent agreement with that observed in experiments, principally when a realistic spatio-temporal intensity distribution is introduced into the calculations. The influence of the coherence of the light on resonant MPI is discussed within the same calculation model. It is found that the shifts of the resonance peaks observed with a chaotic light are much larger than those corresponding to a coherent light of the same averaged intensity. In contrast to what happens in a nonresonant process, the theory predicts, for a near-resonant process, a considerable enhancement of the coherent peaks with respect to the chaotic ones.

To end this discussion we present, in Section V, a method that enables us to calculate exactly the matrix elements of the operator representing the sum of the perturbation series. It is based on the generalization of the technique used in LOPT to an all-order theory, and is well adapted to problems dealing with several continua and to processes induced by very intense radiation fields.

II. THEORETICAL BACKGROUND

Our theoretical approach to multiphoton ionization processes involves a fully quantum-mechanical description of the system of atom plus field. Within this framework, the resolvent formation provides a powerful method for consistently apprehending the problems with which we are concerned.

The resolvent operator $G(z)$ (Goldberger and Watson, 1964; Mower, 1966) is defined by

$$G(z) = 1/(z - H) \tag{2.1}$$

where z is a complex number (energy) and H is the total Hamiltonian in the Schrödinger picture

$$H = H^0 + V \tag{2.2}$$

In Eq. (2.2), H^0 is the free Hamiltonian of the atom-plus-field system, i.e., $H^0 = H^{0(AT)} + H^{0(F)}$, and the interaction V is written in the dipole approximation as

$$V = \mathbf{E}(0)\mathbf{r} \tag{2.3}$$

where $\mathbf{r}$ is the coordinate of the optical electron and $\mathbf{E}(0)$, the electric field at time $t = 0$, is given by

$$\mathbf{E}(0) = i\left(\frac{2\pi}{L^3}E_p\right)^{1/2}(\varepsilon a - \varepsilon^* a^+) \tag{2.4}$$

where E_p is the photon energy; ε is the complex ionization vector of the field; a^+ and a are the creation and destruction operators of a photon, respectively; and L^3 is the quantization volume. It is convenient to put the interaction operator V in the following form

$$V = V^+ + V^- \tag{2.5}$$

where V^- and V^+ are the absorption and emission operators of a photon, respectively. In addition, we shall often be concerned with the dipole operator D defined through the relations $V^- = aD$ and $V^+ = a^+D^*$. The expressions of the operators $V^\pm$ and D are easily obtained from Eqs. (2.3) and (2.4).

From the resolvent operator of Eq. (2.1) we determine the evolution operator $U(t)$ through the integral

$$U(t) = \frac{1}{2\pi i} \oint G(z)e^{-izt}\, dz \tag{2.6}$$

which is performed by using the residue theorem once the integration contour has been suitably chosen (Mower, 1966; Lambropoulos, 1976).

The transition amplitude of any process is defined as the matrix element

$$U_{ba}(t) = \langle b| U(t)|a \rangle \tag{2.7}$$

where $|b\rangle$ and $|a\rangle$ are eigenstates of H^0. For N-photon ionization we have $|b\rangle \equiv |f\rangle \otimes |n - N\rangle$ and $|a\rangle \equiv |g\rangle \otimes |n\rangle$, where $|g\rangle$ and $|f\rangle$ are the atomic states and $|n - N\rangle$ and $|n\rangle$ are the field states expressed within the occupation number representation.

The probability of finding the system in the state $|b\rangle$ at time t is

$$P(t) = |U_{ba}(t)|^2. \tag{2.8}$$

Equations (2.6)–(2.8) outline the importance of knowing the matrix elements of $G(z)$. To calculate this matrix element we would invert the matrix $z - H$. In general, this cannot be done since H cannot be diagonalized in the whole space of states. To tackle this difficulty we can make a perturbative treatment of $G(z)$. By the well-known Feynman identity, Eq. (2.1) yields

$$G = G^0 + G^0VG \tag{2.9}$$

where $G^0 = (z - H^0)^{-1}$ is the resolvent of H^0. This equation can be solved by iteration and $G(z)$ is expanded as a power series in V

$$G = G^0 + G^0VG^0 + G^0VG^0VG^0 + \cdots \tag{2.10}$$

The terms contributing to a definite process are isolated in Eq. (2.10) from a suitable choice of the initial and final states of the system.

For example, for two-photon absorption $G(z)$ is expressed by the infinite series

$$G = G^0 V^- G^0 V^- G^0 + G^0 V^+ G^0 V^- G^0 V^- G^0 V^- G^0$$
$$+ G^0 V^- G^0 V^+ G^0 V^- G^0 V^- G^0 + \cdots \qquad (2.11)$$

the terms of which are obtained by considering all the possible combinations of an arbitrary number of absorptions and emissions of photons in such a way that the net number of absorptions remains equal to two. At moderate intensity and far from any resonance, only the term containing two absorption operators is retained in Eq. (2.11). The other contributions are neglected because they are of higher order. In doing this approximation, $G(z)$ is worked out within the lowest-order perturbation theory (LOPT).

For a transition taking place between the states $|a\rangle = |g\rangle \otimes |n\rangle$ and $|b\rangle = |f\rangle \otimes |n - z\rangle$, the poles that are to be taken into account are $E_a = E_g + nE_p$ and $E_b = E_f + (n - 2)E_p$. Therefore, by expanding G^0 on the basis of the atomic states

$$G^0 \rightarrow \sum_i \frac{|i\rangle\langle i|}{z - E_i - H^{0(F)}} \qquad (2.12)$$

one obtains from Eqs. (2.4), (2.6), and (2.11) the familiar expression of the probability per unit time for 2-photon absorption

$$W_{fg}^{(2)} = \lim_{t \to \infty} P_{fg}^{(2)}(t)/t$$

$$= 2\pi \left(\frac{F}{F_0} E_p \right)^2 \left| \sum_i \frac{(\mathbf{r} \cdot \boldsymbol{\varepsilon})_{fi} (\mathbf{r} \cdot \boldsymbol{\varepsilon})_{ig}}{E_g + E_p - E_i} \right|^2 \delta(E_g + 2E_p - E_f) \qquad (2.13)$$

where F is the photon flux in $cm^{-2} s^{-1}$ and $F_0 = 3.22 \times 10^{34} cm^{-2} s^{-1}$.

A straightforward generalization of Eq. (2.13) shows that the N-photon transition rate varies as the Nth power of the incoming radiation flux

$$W_{fg}^{(N)} = 2\pi \left(\frac{F}{F_0} E_p \right)^N |M_{fg}^{(N)}|^2 \delta(E_g + NE_p - E_f) \qquad (2.14)$$

where

$$M_{fg}^{(N)} = \sum_{i_1} \sum_{i_2} \cdots$$

$$\times \sum_{i_{N-1}} \frac{(\mathbf{r} \cdot \boldsymbol{\varepsilon})_{gi_1} (\mathbf{r} \cdot \boldsymbol{\varepsilon})_{i_1 i_2} \cdots (\mathbf{r} \cdot \boldsymbol{\varepsilon})_{i_{N-1} f}}{(E_g + E_p - E_{i_1})(E_g + 2E_p - E_{i_2}) \cdots (E_g + (N-1)E_p - E_{i_{N-1}})} \qquad (2.15)$$

The alternative to calculating the matrix elements of $G(z)$ involves the inversion of the matrix $z - H$ in a subspace ε spanned by few particular states which are expected to give the most important contributions. According to a well-known technique (Goldberger and Watson, 1964; Mower, 1966; Cohen-Tannoudji, 1967), one defines P and Q as the projection operators onto and outside of this subspace, respectively. They verify the obvious projector relations $P + Q = 1$, $P^2 = P$, $Q^2 = Q$, $PQ = QP = 0$.

As a result of elementary operator algebra, one finds that the projection of $G(z)$ onto ε can be expressed in terms of the shift operator R as

$$PGP = (z - H_0 - PRP)^{-1} \tag{2.16}$$

where

$$R = V + VQ \frac{1}{z - H_0 - QVQ} QV \tag{2.17}$$

For example, if ε is a two-dimensional space spanned by $|a\rangle$ and $|b\rangle$, respectively the initial and the final states of the system, i.e., $P = |a\rangle\langle a| + |b\rangle\langle b|$, one finds

$$G_{ba} = \frac{R_{ba}}{(z - E_a - R_{aa})(z - E_b - R_{bb}) - R_{ba}R_{ab}} \tag{2.18}$$

which is the matrix element required to calculate the transition probability. From similar arguments one can easily determine the expressions of G_{ab}, G_{aa}, and G_{bb}.

The matrix elements of R that appear in Eq. (2.18) are calculated by expanding the last term in the right-hand side of Eq. (2.17) in the power series in V

$$R = V + V \frac{Q}{z - H^0} V + V \frac{Q}{z - H^0} V \frac{Q}{z - H^0} V + \cdots \tag{2.19}$$

Even in the case where one considers only the first few terms of the expansion (2.19) for R, this method provides a considerable improvement in comparison to the preceding one (LOPT) because radiative level shifts are self-consistently taken into account. As it was pointed out (Cohen-Tannoudji, 1967), this corresponds to a re-summation of the perturbation series over a particular class of diagrams.

In the presence of a resonance, or for very intense radiation fields, it is not possible to truncate the series (2.19) because it diverges. Therefore one must take into account all the terms contributing to a well-defined matrix element of R. As a result of an exact re-summation of the series (2.19) (Gontier et al., 1975, 1976), one finds that the diagonal matrix elements R_{aa} and R_{bb} are to be calculated from

$$R^D = V^+ \bar{\tau}_\Box G^0 \bar{V}^- + V^- \bar{\tau}_\Box G^0 \bar{V}^+ \tag{2.20a}$$

while the nondiagonal matrix elements R_{ab} and R_{ba}, corresponding to a transition where $E_b - E_a = NE_p$, are to be calculated from

$$R^{ND} = V^+ \bar{\tau}_\square (G^0 \bar{V}^- \bar{\tau}_0)^{N+1} + V^- \bar{\tau}_\square (G^0 \bar{V}^- \bar{\tau}_0)^{N-1} \qquad (2.20b)$$

and

$$R^{ND} = V^- \bar{\tau}_\square (G^0 \bar{V}^+ \bar{\tau}_\emptyset)^{N+1} + V^+ \bar{\tau}_\square (G^0 \bar{V}^+ \bar{\tau}_\emptyset)^{N-1} \qquad (2.20c)$$

respectively.

In Eq. (2.20) the operators $\bar{\tau}$ are continued fractions of $\bar{V}^+ = QV^+$ and $\bar{V}^- = QV^-$, where

$$\bar{\tau}_0 = \frac{1}{z - H^0 - \bar{V}^- \bar{\tau}_0 \bar{V}^+} \qquad (2.21a)$$

$$\tau_\emptyset = \frac{1}{z - H^0 - \bar{V}^+ \bar{\tau}_\emptyset \bar{V}^-} \qquad (2.21b)$$

$$\bar{\tau}_\square = \frac{1}{z - H^0 - \bar{V}^- \bar{\tau}_0 \bar{V}^+ - \bar{V}^+ \bar{\tau}_\emptyset \bar{V}^-} \qquad (2.21c)$$

III. NONRESONANT IONIZATION

A. Matrix Formulation

As was previously pointed out, nonresonant ionization occurring at moderate intensity provides a typical example where the lowest-order perturbation theory gives satisfactory results. Within this framework, one of the basic difficulties encountered in making a quantitative analysis lies in the calculation of high-order transition matrix elements.

Our aim is not to present an exhaustive account of the works dealing with this problem because many review papers have been published elsewhere (Bakos, 1974; Lambropoulos, 1976; Georges and Lambropoulos, 1980; Morellec et al., 1982). We only mention that many calculation methods have been proposed since the pioneer work of Göppert-Mayer (1931) on 2-photon absorption. Among these methods that of Bebb and Gold (1966) and Bebb (1967) permitted us to calculate high-order processes in many gases. Although the results they found in hydrogen have since been improved by more sophisticated theories, most of the values of the probabilities obtained in rare gases and alkali suffer the comparison with recent calculations.

The method we intend to discuss in more detail belongs to the ensemble of methods in which the exact value of the transition matrix element can be determined. In contrast to the Green's function method (Klarsfeld, 1969,

1970; Zon et al., 1972; Karule, 1971, 1974; Laplanche et al., 1976; Maquet, 1977) where the probability is expressed in terms of hypergeometric functions, the method we are concerned with consists of solving inhomogeneous differential equations. This technique, which was first proposed by Dalgarno and Lewis (1955) and reformulated later by Schwartz and Tieman (1959), has been used by Zernik (1964), and Zernik and Klopfenstein (1965) to investigate 2-photon ionization of hydrogen in the 2s metastable state.

In this section we show how this method has been generalized to perturbation orders higher than two (Gontier and Trahin, 1968, 1971, 1973).

For clarity, $G(z)$ is determined to lowest-order from the expansion of Eq. (2.10). By using well-known relations concerning the field operators, i.e., $a|n\rangle = \sqrt{n}|n - 1\rangle$, $a^{+}|n\rangle = \sqrt{n + 1}|n + 1\rangle$, the matrix element representing N-photon absorption can be written as

$$G_{fg}^{(N)} = \frac{[n(n - 1) \cdots (n - N)]^{1/2}}{(z - E_a)(z - E_b)} M_{fg}^{(N)} \tag{3.1}$$

where

$$M_{fg}^{(N)} = \langle f|D^{(n-N+1)}G^0 D^{(n-N+2)}G^0 \cdots {}^{(n-2)}G^0 D^{(n-1)}G^0 D|g\rangle \tag{3.2}$$

where ${}^{(k)}G^0 = \langle k|G^0|k\rangle$ denotes the average of G^0 over the photon state $|k\rangle$. Since in this case G^0 is diagonal, its average can be expressed in terms of ${}^{(k)}H^0 = z - H^{0(AT)} - kE_p$ and the average of H^0 can be expressed as

$$^{(k)}G^0 = 1/(z - {}^{(k)}H^0) \tag{3.3}$$

For the sake of compactness we adopt a matrix formulation which will prove to be well adapted to more complicated problems. The basic argument is that the operators ${}^{(k)}H^0(r)$ and $D(r)$, as well as the initial and the final states, can be expressed as block matrices and block vectors in the basis of the eigenstates $Y_{lm}(\theta\varphi)$ of the angular momentum operator L^2. Thus, each matrix element is labeled by two pairs of indices representing the values of the orbital and the magnetic quantum numbers. Such operators are called tetradics (Zwanzig, 1960, 1964).

Within this representation ${}^{(k)}H^0$ is a diagonal differential operator with respect to the variable r. For hydrogen we have

$$^{(k)}H^0_{ll'mm'}(r) = \left[-\frac{d^2}{dr^2} - \frac{2}{r} + \frac{l'(l' + 1)}{r^2} + kE_p \right] \delta_{ll'}\delta_{mm'} \tag{3.4}$$

This operator acts on the matrix elements of D (which are functions of r). The form of the matrix D depends on the dipole selection rules on l and m as dictated by the polarization of the light. For linearly polarized light we have $\Delta l = \pm 1$, $\Delta m = 0$, while for right and left polarization one has $\Delta l = \pm 1$, $\Delta m = +1$ and $\Delta l = \pm 1$, $\Delta m = -1$, respectively.

For example, the matrix element of D corresponding to linearly polarized light is

$$D_{ll'mm'}(r) = r[A(l', m')\delta_{ll'-1} + A(l' + 1, m')\delta_{ll'+1}]\delta_{mm'} \tag{3.5}$$

where (Bethe and Salpeter, 1957)

$$A(L, M) = \left[\frac{(L + M)(L - M)}{(2L + 1)(2L - 1)}\right]^{1/2} \tag{3.6}$$

Similarly, one could determine the matrix element of D for circular polarization. One would find

$$D_{ll'mm'}(r) = r[B(l' + 1, m' + 1)\delta_{ll'+1} - C(l', m')\delta_{ll'-1}]\delta_{mm'+1} \tag{3.7}$$

and

$$D_{ll'mm'}(r) = r[-C(l' + 1, m' - 1)\delta_{ll'+1} + B(l', m')\delta_{ll'-1}]\delta_{mm'-1} \tag{3.8}$$

for right and left circularly polarized light, respectively. The angular coefficients are given by

$$B(L, M) = \left[\frac{(L + M)(L + M - 1)}{(2L + 1)(2L - 1)}\right]^{1/2} \tag{3.9}$$

and

$$C(L, M) = \left[\frac{(L - M)(L - M - 1)}{(2L + 1)(2L - 1)}\right]^{1/2} \tag{3.10}$$

B. Calculation Technique

To calculate $M_{fg}^{(N)}$, we define the following ket operators

$$|W^{(n-N+1)}\rangle = {}^{(n-N+1)}G^0 D {}^{(n-N+2)}G^0 \cdots {}^{(n-2)}G^0 D {}^{(n-1)}G^0 D |g\rangle \tag{3.11a}$$

$$|W^{(n-N+2)}\rangle = {}^{(n-N+2)}G^0 \cdots {}^{(n-2)}G^0 D {}^{(n-1)}G^0 D |g\rangle \tag{3.11b}$$

$$|W^{(n-2)}\rangle = {}^{(n-2)}G^0 D {}^{(n-1)}G^0 D |g\rangle \tag{3.11c}$$

$$|W^{(n-1)}\rangle = {}^{(n-1)}G^0 D |g\rangle \tag{3.11d}$$

By applying the operators $(z - {}^{(n-N+1)}H^0)$, $(z - {}^{(n-N+2)}H^0)$, ..., $(z - {}^{(n-2)}H^0)$ and $(z - {}^{(n-1)}H^0)$ to both sides of Eqs. (3.11a), (3.11b), ..., (3.11c), and (3.11d), respectively, one obtains the following set of operator equations:

$$(z - {}^{(n-N+1)}H^0)|W^{(n-N+1)}\rangle = D|W^{(n-N+2)}\rangle \tag{3.12a}$$

$$(z - {}^{(n-N+2)}H^0)|W^{(n-N+1)}\rangle = D|W^{(n-N+3)}\rangle \tag{3.12b}$$

$$(z - {}^{(n-2)}H^0)|W^{(n-2)}\rangle = D|W^{(n-1)}\rangle \tag{3.12c}$$

$$(z - {}^{(n-1)}H^0)|W^{(n-1)}\rangle = D|g\rangle \tag{3.12d}$$

The sets of equations to be solved are obtained by expressing each operator within the preceding matrix representation. In this way one straightforwardly obtains all the contributing orbital channels.

The strategy for solving the set of equations (3.12a)–(3.12d) is transparent. One determines $|W^{(n-1)}\rangle$ by solving Eq. (3.12d). This solution is substituted in the right-hand side of Eq. (3.12c) from which one determines $|W^{(n-2)}\rangle$, etc. The solution $|W^{(n-N+1)}\rangle$ of the last equation is used to calculate the required matrix element which is given by

$$M_{fg}^{(N)} = \langle f|D|W^{(n-N+1)}\rangle \tag{3.13}$$

Equations (3.12a)–(3.12d) summarize what is called the implicit summation technique. Obviously it involves a coordinate representation of $H^{0(AT)}$ rather

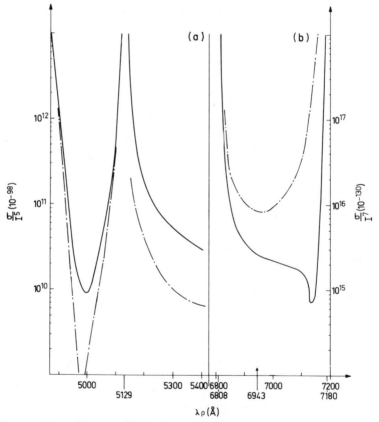

Fig. 1. (a) Dispersion rate σ_I/I^5 for 6-photon ionization of H in the ground state, $N = 6$, $n = 1$; (b) dispersion rate σ_I/I^7 for 8-photon ionization of H in the ground state, $N = 8$, $n = 1$. Present calculations, solid lines; data calculated by Bebb and Gold (1966), dashed lines.

than its usual energy representation. This can be used in complex atoms. For that purpose one replaces the hydrogen potential $1/r$ in Eq. (3.4) by $V(r)$ the relevant atomic potential (Crance and Aymar, 1979; Aymar and Crance, 1979).

Our interest in using such a technique rather than an appproximate one (Bebb and Gold, 1966) is apparent in Fig. 1, where 6- and 8-photon ionization rates of hydrogen are represented.

We note that an important discrepancy appears between two resonances. This is because in this region no particular state prevails. It is suitable to use a calculation technique which takes exactly into account the complete set of atomic states.

The minima of the dispersion curves are sometimes interpreted in terms of destruction interferences resulting from contributions of different parts of the spectrum. Let us consider the Nth-order transition matrix element expressed in the energy representation as given by Eq. (2.15). The energy denominator $E_g + (N - s)E_p + E_{i_{N-s}}$ will be less than 0 or greater than 0 according to whether $|E_{i_{N-s}}|$ is less than $|E_g + (N - s)E_p|$ or not. Thus in the sum over i_{N-s}, the amplitudes of opposite signs combine with each other to give, after squaring M_{fg}, negative contributions to the probability that are similar to the interference terms encountered in optics.

IV. RESONANT IONIZATION

A. The Method of Projectors

According to a well-known definition encountered in many domains of physics, a resonance appears in atomic processes induced by intense radiation fields when any proper frequency of the atom coincides with a multiple of the frequency of the exciting field. From Eq. (2.15) it is clear that the atom behaves like a resonator whose frequencies are $(E_g - E_{i_k})/\hbar$, and that the driving frequencies are integer multiples of the frequency of the field $E_p/\hbar$. Thus at resonance we have $E_g - E_{i_k} \pm kE_p$. As a consequence, the corresponding energy denominator tends to zero, while the transition probability given by Eqs. (2.14) and (2.15) becomes infinite. Such a behavior indicates that LOPT does not provide a realistic description of resonant processes. The reason is that the higher-order terms mentioned in Section II give nonnegligible contributions, which must be taken into account. It is thus necessary to elaborate a re-summed theory where all the contributions of the process we consider are treated on an equal footing. To this end one uses a technique sketched in Section II. For a single resonance of order k, i.e., produced by the kth absorbed photon, one calculates the matrix elements of $G(z)$

by inverting the operator $z - H$ in a subspace spanned by $|a\rangle = |g\rangle \otimes |u\rangle$, $|b\rangle = |f\rangle \otimes |n - N\rangle$, and $|c\rangle = |r\rangle \otimes |n - k\rangle$, the initial, final, and resonant states, respectively. The additional (resonant) state considerably lengthens the analytic formulation of the problem. For the sake of compactness we resort to a formation leading to expressions formally identical to those found in the two-level theory. In this respect one considers that the subspace ε defined in Section II can be split into two subspaces. Let P' and Q' be the projectors onto and outside one of these two subspaces η. We find, in terms of a straightforward operator algebra, that

$$P'GP' = \frac{1}{z - H_0 - P'\mathbb{R}P'} \tag{4.1}$$

where

$$\mathbb{R} = R + RQ' \frac{1}{z - H_0 - Q'RQ'} Q'R \tag{4.2}$$

The subspace η itself can be split into two subspaces, etc. In doing this many times, the projection of $G(z)$ onto a particular subspace is of the form shown in Eq. (2.16), where R is to be replaced by an expression determined through recurrent relations like that of Eq. (4.2). This procedure is particularly convenient if one deals with one or several resonant intermediate states.

For a single resonance occurring in multiphoton ionization, η is a two-dimensional space spanned by the bound initial and resonant states $|a\rangle$ and $|c\rangle$, whereas Q' is the projector onto the one-dimensional space containing the continuum state $|b\rangle$, i.e., $Q' = \int dE_b |b\rangle\langle b|$.

From Eq. (4.1) one finds

$$G_{aa} = \frac{z - E_c - \mathbb{R}_{cc}}{\mathbb{D}} \tag{4.3a}$$

$$G_{ca} = \mathbb{R}_{ca}/\mathbb{D} \tag{4.3b}$$

where $\mathbb{D}$ is the determinant of the matrix $z - H_0 - P'RP'$

$$\mathbb{D} = (z - E_a - \mathbb{R}_{aa})(z - E_c - \mathbb{R}_{cc}) - \mathbb{R}_{ac}\mathbb{R}_{ca} \tag{4.4}$$

and

$$\mathbb{R} = R(z) + R(z) \int \frac{|b\rangle\langle b|}{z - E_b - \mathbb{R}_{bb}(z)} dE_b R(z) \tag{4.5}$$

$R(z)$ being given by the re-summed expressions of Eqs. (2.20a)–(2.20c). The matrix elements (4.3) serve to calculate the ionization probability through the expression

$$P(t) = 1 - |U_{aa}(t)|^2 - |U_{ca}(t)|^2 \qquad (4.6)$$

which is simply derived from probability conservation. Thus the mechanism of multiphoton ionization of atoms can be discussed within the model of a two-level system with losses (Shore and Ackerhalt, 1977).

The calculation of $U_{aa}(t)$ and $U_{ca}(t)$ in Eq. (4.6) requires the knowledge of the poles of G_{aa} and G_{ca}. These poles, which are the roots of the equation $\mathbb{D} = 0$, can only be determined numerically because the $\mathbb{R}$ are complicated functions of z. The effect of the coupling with the continuum in Eq. (4.5) is to make complex the matrix elements of R. By assuming that $z = E - i\Gamma$, one can write

$$\mathbb{R}_{ij}(z) = \Delta_{ij}(E, \Gamma) - i\Gamma_{ij}(E, \Gamma) \qquad (4.7)$$

From Eq. (4.7) one can calculate the proper shifts Δ_{aa} and Δ_{cc} and the proper widths Γ_{aa} and Γ_{cc} of the levels a and c.

B. Level Crossing and Anticrossing

The poles of the resolvent or the energy of the dressed atom (Cohen-Tannoudji, 1967; Cohen-Tannoudji and Haroche, 1969) obtained by solving Eq. (4.4) can be written in the form

$$z^{\pm} = E^{\pm}(E_{\mathrm{p}}, I) - i\Gamma^{\pm}(E_{\mathrm{p}}, I) \qquad (4.8)$$

where $E^{\pm}$ and $\Gamma^{\pm}$ are complicated functions of the shifts and the widths of the initial and the resonant levels (Beers and Armstrong, 1975; Gontier and Trahin, 1979a). These expressions predict quite different behaviors of $E^{\pm}$ and $\Gamma^{\pm}$, depending on the relative values of $|\mathbb{R}_{ac}|^2$ and $\frac{1}{4}\Gamma_{cc}^2$, the coupling of the resonant state with the ground state and with the continuum, respectively.

It can be readily verified from Eqs. (4.4), (4.7), and (4.8) that if $|\mathbb{R}_{ca}|^2 > \frac{1}{4}\Gamma_{cc}^2$, the curves representing the values of E^+ and E^- as functions of the intensity never cross each other. Their behavior is shown in Fig. 2a for 3-photon ionization of Cs.

Near the dynamical resonance, i.e., $E_a + \Delta_{aa}(I) \simeq E_c + \Delta_{cc}(I)$, they can be approximated by hyperbolas whose asymptotes are $E_a + \Delta_{aa}(I)$ and $E_c + \Delta_{cc}(I)$, the shifts Δ_{aa} and Δ_{cc} being roughly linear in intensity. For this reason, one says that the level shows an anticrossing.

The situation is quite different if $|\mathbb{R}_{ac}|^2 > \frac{1}{4}\Gamma_{cc}^2$. In this case $E^+ = E^-$ and one is faced with a level crossing. Such a phenomenon is illustrated in

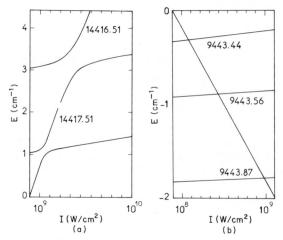

Fig. 2. Energy of the dressed atom as a function of the intensity for (a) a second-order resonance on the $9D_{3/2}$ level in 3-photon ionization and (b) a third-order resonance on the $6F_{5/2}$ level in a 4-photon ionization of Cs.

Fig. 2b for 4-photon ionization of Cs. For the sake of homogeneity, we have adopted the denomination used in the radio-frequency range (Cohen-Tannoudji, 1967).

C. The Ionization Probability

1. Analytic Expression

From the knowledge of the poles $z^{\pm}$, one can calculate the matrix elements of the time evolution operator in Eq. (4.6) by applying the residue theorem to the inversion integral (2.6). One finds that the probability can be expressed in terms of hyperbolic and circular functions of time as

$$P(t) = 1 - \frac{e^{-(\Gamma^+ + \Gamma^-)t}}{(E^+ - E^-)^2 + (\Gamma^+ - \Gamma^-)^2} [a \sinh(\Gamma^+ - \Gamma^-)t + b \sinh(\Gamma^+ - c_1 \cos(E^+ - E^-)t - c_2 \sinh(E^+ - E^-)t] \quad (4.9)$$

where the coefficients a, b, c_1, and c_2 are slowly varying functions of the photon energy (Gontier and Trahin, 1980).

By a simple change in the notation, one finds that this expression for the probability is formally equivalent to that obtained by Beers and Armstrong (1975) within the effective operator formalism. But as a result of our re-summation technique, all the quantities appearing in Eq. (4.9) are calculated to arbitrary order in intensity. Such an all-order theory will prove to be useful for interpreting experimental results obtained with ultra-intense radiation fields. The probability thus calculated concerns the idealistic situation where the amplitude of the radiation field does not show any time-and-space variations, i.e., "square pulse" and isotropic intensity distribution. Since the lasers deliver their energy according to well-defined spatial and temporal distributions, it is not possible to reach the experimental value of the probability. Thus the only reliable comparisons between theory and experiment must involve measured quantities like the ion signal. In this respect all the theoretical results concerning the ions, presented in this paper, are obtained within the collection geometry and the temporal pulse shape observed in resonant 4-photon ionization of Cs (Petite *et al.*, 1979).

On its own, the probability (4.8) supplies enough information about the behavior of the system when the photon frequency, the intensity, and the interaction time are varied. Thus the discussion will refer alternatively to the probability or to the ion number.

2. Frequency and Intensity Dependences

An important consequence of the all-order calculation model we use is that the value of the probability is not infinite at exact resonance. The couplings of the resonant level with the continuum or with other discrete levels introduce an important damping which limits the growth of the probability. This is illustrated in Fig. 3, where the dispersion curves corresponding to resonant 4-photon ionization of Cs for different values of the intensity are represented.

The maxima are well resolved for the two components of the resonant 6F doublet.

In contrast to what happens in LOPT, the position and the width of the resonance peaks depend on the intensity.

The calculated and the measured values of the resonance peak are plotted in Fig. 4 as functions of the intensity.

The crosses are the experimental data corresponding to picosecond experiments (Lompré *et al.*, 1978), and the triangles are the results obtained with a pulse duration of 37 ns (Morellec *et al.*, 1976). We note the excellent agreement between theory and experiment within the whole range of interaction time considered.

The other important effect resulting from higher-order terms concerns the intensity dependence of the probability or the ion number on both sides of the static resonance. This is easily seen by varying the intensity for each chosen photon frequency.

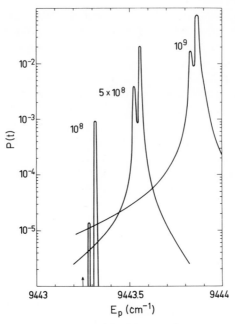

Fig. 3. Four-photon ionization probability of Cs as a function of the photon energy. The intensity is written beside the curves. The interaction time τ is 50 ns and the arrow indicates the position of static resonance on the $6F_{5/2}$ level ($E_p = 9443.254$ cm^{-1}).

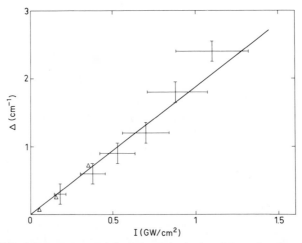

Fig. 4. Shift of the resonance peak for 4-photon ionization of Cs as a function of the intensity. The solid line represents the results obtained from the present work. The experimental data are those of Morellec et al. (1976) (triangles) and of Lompré et al. (1978) (crosses).

One obtains on the curves representing the ion number as functions of the intensity, a distortion denoting the presence of the dynamical resonance arising when the static detuning is exactly balanced by the intensity-dependent level shifts. The smooth bumps observed in Fig. 5 come from this resonance. They appear only on one side of the resonance (negative detunings) and originate a strong asymmetry regarding the order of nonlinearity. This quantity is defined by

$$K = k \log(N_i)/[d \log(I)] \tag{4.10}$$

and is determined from the slopes of the curves of Fig. 5 by keeping constant the intensity or the ion number.

This last procedure is extensively used in experiments because it provides a much large variation range of K.

As long as LOPT holds (nonresonant processes), the probability or the ion number varies according to a simple power law in intensity. Thus, from Eq. (2.14), the intensity dependence of $P(t)$ is of the form I^N. In this case K is the number of absorbed photons ($K = N$). In the presence of a resonance,

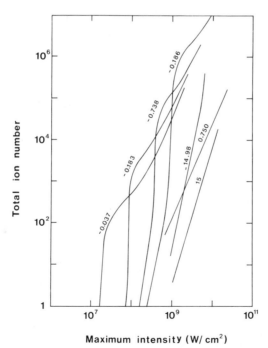

Maximum intensity (W/ cm²)

Fig. 5. Four-photon ionization of Cs. Total ion number as a function of the amplitude of the Gaussian pulse. The interaction time is 40 ns and the number beside the curves is the resonance detuning in cm⁻¹ (static resonance: $E_p = 9443.254$ cm⁻¹).

one observes important deviations from this simple law. As it is shown in Fig. 6, the curves are strongly assymetric with regard to $K = 4$ as one crosses the static resonance. Within a small energy interval the lowest value of K is about 2, while its upper bound is greater than 30. Such features, which are general, can be easily explained from the behavior of the poles of $G(z)$ near resonance (Gontier and Trahin, 1979a). Here again the good agreement between theory and experiment, which is observed on the curve referring to the ion number, confirms the relevance of the calculation model.

3. Time Dependence

Regarding the time dependence of Eq. (4.8), some particular behaviors of the probability can be predicted for special values of the arguments of the functions.

If the resonances are sharp enough and/or for short pulse durations, i.e., $< 10^2$ ps in 4-photon ionization of Cs, one has

$$(E^+ - E^-)t \ll 1 \tag{4.11a}$$

and

$$(\Gamma^+ - \Gamma^-)t \ll 1 \tag{4.11b}$$

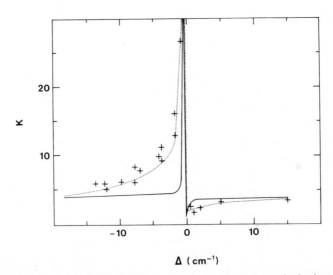

Fig. 6. Order of nonlinearity $K = d \log N_i / d \log I_{max}$ for 4-photon ionization of Cs as a function of the resonance detuning. The solid curve is obtained by keeping constant the amplitude of the Gaussian pulse ($I = 10^7$ W cm^{-2}). The dotted line is determined by prescribing a constant ion number ($N_i = 10^3$). The crosses represent the experimental data of Morellec et al. (1976).

The probability can be expanded in power series in t to give

$$P(t) = \mathbb{R}_{ac}\mathbb{R}_{ca}\Gamma_{cc}\left(\frac{2}{3}t^3 - \frac{1}{2}\Gamma_{cc}t^4 - 4(\tilde{E}_a - \tilde{E}_c + 4\mathbb{R}_{ac}\mathbb{R}_{ca} - 7\Gamma_{cc}^2)\frac{t^5}{5!} + \cdots\right)$$

$$(4.12)$$

where $\mathbb{R}_{ac}$ is the bound–bound matrix element that couples the levels a and c, whose displaced energies are $\tilde{E}_a = E_a + \Delta_{aa}$ and $\tilde{E}_c = E_c + \Delta_{cc}$, respectively. For this singly resonant process, the probability varies as t^3, while the resonance width is proportional inversely to t, i.e., $\gamma_{cm^{-1}} = 33.29/(N-1)t_{ps}$ for a resonance of order $N-1$ in Cs. This result concerning the time dependence of the probability has also been found by Fedorov and Kazakov (1977) and reformulated more recently by McClean and Swain (1978).

As the interaction time increases, a different situation arises. Since near resonance $\Gamma^- \gg \Gamma^+$, the terms containing Γ^- in the exponents are strongly damped and can be neglected. In this case the expression of the probability has the following simple form

$$P(t) = 1 - e^{-2\Gamma^+t}$$

$$(4.13)$$

Providing that the exponent is less than unity, the exponential can be expanded in power series whose leading term shows a linear time dependence. The probability reduces to the following familiar form:

$$P(t) = \frac{\mathbb{R}_{ac}\mathbb{R}_{ca}\Gamma_{cc}}{(\tilde{E}_a - \tilde{E}_c)^2 + \Gamma_{cc}^2}t$$

$$(4.14a)$$

for a level crossing and to

$$P(t) = \frac{\mathbb{R}_{ac}\mathbb{R}_{ca}\Gamma_{cc}}{(\tilde{E}_a - \tilde{E}_c)^2 + 4\mathbb{R}_{ac}\mathbb{R}_{ca}}t$$

$$(4.14b)$$

for a level anticrossing. The corresponding resonance widths are $\gamma = \frac{2}{3}\Gamma_{cc}$ and $\gamma = 2\mathbb{R}_{ac}$, respectively, and are no longer governed by the interaction time. The predictions deduced from power series expressions are corroborated by the results of the computations shown in Figs. 7 and 8, which refer to 4-photon ionization of Cs.

In Fig. 7 the cubic and the linear variations of the probability take place within time intervals usually considered in picosecond and nanosecond experiments, while the domain of linear variations increase as one deviates from exact resonance. In this region the behavior of the process can be described by a time-independent transition rate. The same phenomenon is observed on the total ion signal (dashed curves).

Because of the experimental difficulty of continuously varying the interaction time, such a behavior of the probability has not been observed. Nevertheless, a comparison regarding the resonance width is possible for some typical values of the interaction time.

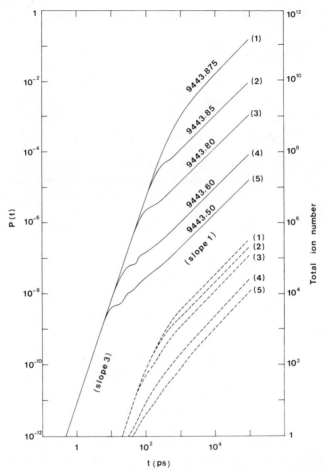

Fig. 7. Four-photon ionization of Cs. The probability calculated with a square pulse at 10^9 W cm^{-2} is plotted against the interaction time for some photon energies in cm^{-1} (solid curves and left-hand side scale). The total ion number obtained with a Gaussian pulse is represented as a function of the width of the Gaussian for some values of E_p (dashed curves and right-hand scale). The digits in parentheses signify one-to-one correspondence between the upper and the lower set of curves.

From the dispersion curves of Fig. 8, whose considerable damping in the picosecond range denote the drastic effect of the pulse duration, one determines the widths, which, from Table I, are in pretty good agreement with the corresponding experimental values.

Before ending this discussion devoted to the influence of the interaction time on the probability, we note that from the dispersion curves of Fig. 3 it

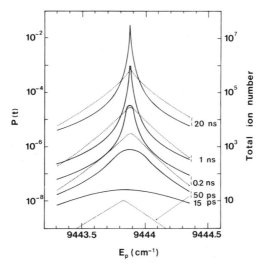

Fig. 8. Dispersion curves of 4-photon ionization of Cs for different interaction times. The solid curves (left-hand-side scale) represent the probability calculated with a square pulse of amplitude $I = 10^9$ W cm^{-2}. The dotted curves right-hand-side scale show the total ion number produced by Gaussian pulses of same duration (at half-height) and of equal amplitude, within the experimental device of Morellec et al. (1976) (resonance: $E_p = 9443.254$ cm^{-1}).

is possible to define a quantity which permits additional comparisons between theory and experiments. To this end one plots the amplitude of the resonance peaks against the laser intensity in a log–log scale. In this representation, the points are distributed on a curve whose slope depends on the intensity and the interaction time. In Fig. 9 the solid lines represent the theoretical variations in this slope as a function of the interaction time for

Table I

Four-Photon Ionization of Cs[a]

t (ps)	γ_L (cm^{-1})	γ_{exp} (cm^{-1})	γ_{th} (cm^{-1})
15	1.4	1 ± 0.1	0.88
50	0.4	$0.3 \pm 5(-2)$	0.29
1.5(3)	2(−2)	$0.2 \pm 5(-2)$	0.12

[a] Experimental (γ_{exp}) and theoretical (γ_{th}) resonance widths corresponding to the two components of the 6F resonant doublet. The laser bandwidth γ_L and the number in parentheses indicate the powers of 10. The experimental values of the width of the resonance have been obtained at 10^9, 5×10^8, and 10^7 W cm^{-2}, respectively, whereas γ_{th} has been computed at 10^9 W cm^{-2}.

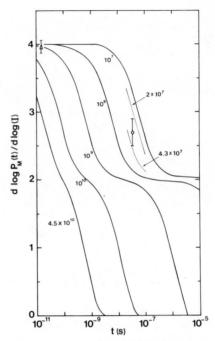

Fig. 9. Variations of the slope $K(I, t) = d \log P_M(t)/d \log (I)$ as a function of time for resonant 4-photon ionization of cesium. For the clearness of the figure of curves corresponding to $I = 2 \times 10^7$ and 4.3×10^7 W cm^{-2} have only been reported in the region of interest. The experimental results obtained with nanosecond and picosecond pulses are represented by a circle (Morellec *et al.*, 1979) and a triangle (Lompré *et al.*, 1978), respectively.

different intensity values. The three variation regimes exhibited by the curves of Fig. 9 correspond to the typical variations of the probability previously discussed. For small interaction time, the probability varies as the fourth power of the intensity and is proportional to t^3 (upper plateaus). As the time increases, the probability becomes linear in time and quadratic in intensity (intermediate plateaus). Later one reaches the saturation. The triangle and the circle represent the values of the slope measured in the picosecond (Lompré *et al.*, 1978) and in the nanosecond (Morellec *et al.*, 1979) experiments, respectively.

At 15 ps the point is centered on the curve 10^9 W cm^{-2}, which is the averaged intensity in this experiment.

At 37 ns its value is bounded by those corresponding to experimental intensity extrema. In both experiments the values of the slope are the ones predicted by the calculations. From the knowledge of this slope and the interaction time one can determine the intensity.

D. Photon Statistics

The effects of photon statistics on nonresonant multiphoton processes have been extensively discussed during the past few years (Glauber, 1963, 1965; Lambropoulos et al., 1966; Guccione-Gush et al., 1967; Lambropoulos, 1968; Mollow, 1968; Debethune, 1972). The general prediction of these calculations is that the N-photon transition probability induced by a single mode thermal light is enhanced by a factor $N!$ with regard to that corresponding to a coherent light of equal average photon number. Such coherent effects have been observed experimentally in 11-photon ionization of Xe (Lecompte et al., 1974, 1975). The situation can be quite different in the presence of a resonance where the Gaussian intensity distribution can affect the slope of the resonance peak (Armstrong et al., 1976; Mostowski, 1976). In fact, the amplitude, the width, and the position of the resonance peaks are strongly modified by a change in the photon statistics (Gontier and Trahin, 1979b).

A quantitative analysis can be done within the preceding calculation model. Assuming a P representation for the density matrix of the field, i.e., $\rho_F(0) = \int P(\alpha)|\alpha\rangle\langle\alpha|\,d^2\alpha$ (Glauber, 1963, 1965), the probability is found to be

$$P(t) = \int P(\alpha)\langle\alpha|\,U_{gf}^+(t)U_{fg}(t)|\alpha\rangle\,d^2\alpha \tag{4.15}$$

where $|\alpha\rangle$ is the Glauber's coherent state and $U(t)$ is the time evolution operator. In LOPT the matrix element of $U(t)$ accounting for N-photon ionization factors into atomic and field variables according to

$$U^{(N)}(t) = \sigma^{(N)}(t)(a)^N \tag{4.16}$$

and Eq. (3.14) reduces to

$$P(t) = |\sigma_{fg}^{(N)}(t)|^2 \int P(\alpha)\langle\alpha|(a^+)^N(a)^N|\alpha\rangle\,d^2\alpha \tag{4.17}$$

where the quasi-probability $P(\alpha)$ is given for an ideal laser by

$$P(\alpha) = \frac{1}{2\pi|\alpha|}\,\delta(|\alpha| - |\alpha_0|) \tag{4.18}$$

while for a chaotic field it is

$$P(\alpha) = \frac{1}{\pi|\alpha_0|^2}\exp{-\frac{|\alpha|^2}{|\alpha_0|^2}} \tag{4.19}$$

Since $|\alpha\rangle$ is eigenstate of a, i.e., $a|\alpha\rangle = \alpha|\alpha\rangle$, one finds from Eqs. (4.17), (4.18), and (4.19) that the ratio of the probabilities calculated with a chaotic light to that corresponding to a coherent light is

$$P_{(t)}^{\text{ch}}/P_{(t)}^{\text{c}} = N! \tag{4.20}$$

This well-known result which holds in LOPT, fails when $U^{(N)}(t)$ is a complicated function of the field operators like the one involved in the discussion of resonant processes.

In this case, the average of Eq. (4.15) no longer reduces to that of a simple product of field operators written in a normal form.

To calculate this average one introduces the following representation of the coherent states (Glauber, 1965)

$$|\alpha\rangle = \sum_n \frac{\alpha^n}{n!^{1/2}} \exp -\frac{1}{2}|\alpha\rangle^2|n\rangle \qquad (4.21)$$

which enables us to write the N-photon ionization probability in the form

$$P(t) = \sum_n |U_{fg}^{(N,n)}|^2 \int P(\alpha)e^{-|\alpha|^2|\alpha|^{2n}} d^2\alpha \qquad (4.22)$$

where the superscript n indicates that $U(t)$ is a function of the photon number in the mode.

In considering the expressions of the quasi-probability $P(\alpha)$ given by Eqs. (4.18) and (4.19), one finds that the ionization probabilities corresponding to a coherent and a chaotic light are

$$P^{(c)}(t) = \sum_{n=0}^{\infty} \frac{\bar{n}}{n!} e^{-n}|U_{fg}^{(N,n)}(t)|^2 \qquad (4.23a)$$

and

$$P^{(ch)}(t) = \sum_{n=0}^{\infty} \frac{\bar{n}^n}{(1+\bar{n})^{n+1}} |U_{fg}^{(N,n)}(t)|^2 \qquad (4.23b)$$

respectively, $\bar{n}$ being the averaged photon number in the mode. In Eqs. (4.23a) and (4.23b), the probability is the sum of components, which are infinite in number. Each of them is equal to the ionization probability corresponding to a radiation field containing n photons times the probability for the field to be in such a n-photon state. The atomic response to the excitation is entirely contained in the function $U_{fg}^{(N,n)}$, which is calculated from an accurate method.

The theory is applied to 4-photon ionization of cesium by a single mode field of vanishing linewidth.

From the curves A, B, and C of Fig. 10, one sees that the Gaussian peaks are broadened and shifted with regard to the coherent ones. These effects which come from photon correlations become more and more sensitive when the intensity increases. As a result of a calculation (Zoller, 1979; Lompré et al., 1981) where the sum over n in Eq. (4.23) is replaced by an integral, one finds, if the single rate approximation holds, that the "Gaussian" shift

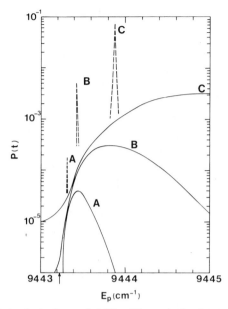

Fig. 10. Four-photon ionization probability of Cs versus the photon energy. The resonance peaks corresponding to a coherent (dashed curves) and a chaotic light (solid curves) are reported for three values of the average intensity [A, 10^8; B, 3×10^8, and C, 10^9)]. The arrow gives the position of the static resonance with the $6F_{5/2}$ state ($E_p = 9443.254$ W cm^{-1}).

must be $N - 1$ times the "coherent" one. In the present case, the ratio of two shifts is roughly equal to 3, i.e., 3, 3.2, and 2.97 for the curves A, B, and C, respectively.

On the other hand, near resonance the coherent light can be more efficient than the Gaussian light. This is the reverse behavior of that exhibited by nonresonant processes. Nevertheless one must mention that the prevalence of the coherent light over the chaotic one is observed in the close neighborhood of the coherent peaks. Since they are extremely peaked the coherent resonance curves are more sensitive to the spatio-temporal energy distribution of the laser than the Gaussian ones. Thus the enhancement predicted by the theory can be considerably reduced. This is observed in experiments (Lompré *et al.*, 1981) where, for example, the Gaussian peak prevails over the coherent ones even at resonance. But the enhancement in favor of the Gaussian light, which is about 3, is much smaller than 4!, which is its value observed far from resonance. In this respect, the behavior of the shifts and the amplitudes of the resonance peaks are those predicted by our zero-bandwidth model.

V. GENERALIZED IMPLICIT SUMMATION TECHNIQUE

In Section III we described a method enabling the exact calculation of Nth-order transition matrix element within LOPT. This technique, which is easily handled when one deals with unperturbed energy denominators, does not apply, in its present form, to re-summed expressions of $G(z)$ and must be generalized to this case.

This generalized implicit summation technique (GIST) (Gontier et al., 1981a,b) can be considered as the last link that allows the exact quantitative analysis of a large variety of problems encountered in this domain of physics to be made. For example, those concerning nonresonant interactions with ultraintense fields or those involving interactions between many continua.

For clarity, the method is exemplified by considering 2-photon absorption. As a result of the re-summation of the perturbation series (Gontier et al., 1976), the matrix element accounting for 2-photon absorption is

$$G_{fg}^{(2)} = \sqrt{n(n-1)}\, M_{fg}^{(2)} \tag{5.1}$$

where, in the notation of Sections II and III, $M_{fg}^{(2)}$ is written as

$$M_{fg}^{(2)} = \langle f| \frac{1}{z - {}^{(n-2)}H_0 - {}^{(n-2)}R^+ - {}^{(n-2)}R^-}$$

$$\times D \frac{1}{z - {}^{(n-1)}H_0 - {}^{(n-1)}R^-} D \frac{1}{z - {}^{(n)}H_0 - {}^{(n)}R^-} |g\rangle \tag{5.2}$$

The continued fractions $R^{\pm}$ being defined by

$$R^{\pm} = V^{\pm} G^{\pm} V^{\pm} \tag{5.3a}$$

and

$$G^{\pm} = \frac{1}{z - H^0 - R^{\pm}} \tag{5.3b}$$

As before, all the operators of Eq. (5.2) are represented by matrices of finite size in the basic of the eigenstates of the angular momentum operator.

According to what was previously done in Eqs. (3.11), we define the kets

$$|V_g^{(n-2)}\rangle = {}^{(n-2)}GD|V_g^{(n-1)}\rangle \tag{5.4a}$$

$$|V_g^{(n-1)}\rangle = {}^{(n-1)}G^- D|V_g^{(n)}\rangle \tag{5.4b}$$

$$|V_g^{(n)}\rangle = {}^{(n)}G^- |g\rangle \tag{5.4c}$$

which lead to the following set of differential equations

$$(z - {}^{(n-2)}H^0)|V_g^{(n-2)}\rangle - {}^{(n-2)}R|V_g^{(n-2)}\rangle = D|V_g^{(n-1)}\rangle \tag{5.5a}$$

$$(z - {}^{(n-1)}H^0)|V_g^{(n-1)}\rangle - {}^{(n-1)}R^-|V_g^{(n-1)}\rangle = D|V_g^{(n)}\rangle \qquad (5.5\text{b})$$

$$(z - {}^{(n)}H^0)|V_g^n\rangle - {}^{(n)}R^-|V_g^{(n)}\rangle = |g\rangle \qquad (5.5\text{c})$$

by applying successively the operators $z - {}^{(n-2)}H^0 - {}^{(n-2)}R$, $z - {}^{(n-1)}H^0 - {}^{(n-1)}R^-$, and $z - {}^{(n)}H^0 - {}^{(n)}R^-$ to the left of $|V_g^{(n-2)}\rangle$, $|V_g^{(n-1)}\rangle$, and $|V_g^{(n)}\rangle$, respectively; G is defined in terms of $R = R^+ + R^-$ as $G = (z - H^0 - R)^{-1}$.

From the knowledge of $|V_g^{(n-2)}\rangle$, one calculates $M_{fg}^{(2)}$ using

$$M_{fg}^{(2)} = \langle f|V_g^{n-2}\rangle \qquad (5.6)$$

The sets of Eqs. (3.12) and (5.5) are formally solved in the same way. In both cases one determines the solution of the last equation, which gives the inhomogeneous terms of the preceding equation, etc. The complication arising from the use of the GIST is that the solution of each of the Eqs. (5.5) is determined itself from a set of coupled equations whose number depends on the size of the matrices representing the operators and on the number of iterations of the operators R and R^-. This is easily seen by iterating once of operators R^- of Eq. (5.5c). One obtains

$$(z - {}^{(n)}H^0)|V_g^{(n)}\rangle - (n + 1)D|Y_g^{(n+1)}\rangle = |g\rangle \qquad (5.7\text{a})$$

$$(z - {}^{(n+1)}H^0)|Y_g^{(n+1)}\rangle - D|V_g^{(n)}\rangle - {}^{(n+1)}R^{-(n+1)}G^-D|V_g^{(n)}\rangle = 0 \qquad (5.7\text{b})$$

where

$$|Y_g^{(n+1)}\rangle = {}^{(n+1)}G^-D|V_g^{(n)}\rangle \qquad (5.8)$$

By iterating the operators R, each Eq. (5.5) is equivalent to a set of coupled equations, which are infinite in number. Each set can be truncated by neglecting the term containing the operator R.

Finally, by replacing the operators by their matrix representation the problem of the double summation (perturbation series and intermediate states) is formulated in a fully tractable form.

VI. CONCLUDING REMARKS

The results presented in this short account are representative of those obtained in terms of more detailed studies. They outline the dependence of multiphoton ionization processes with respect to the field intensity, the interaction time, and the photon statistics. From the results obtained within an accurate calculation model one can draw some general conclusions.

Most comparisons between theory and experiment are done on quantities involving the variations of the probability or the ion signal rather than its absolute value. In this respect, the agreement between the calculated and the measured values is excellent and confirms the relevance of an all-order theory.

Concerning nonresonant ionization, it has been shown that important errors can be introduced in the calculation of the probability, when the sums over the intermediate states are truncated. Some methods enable one to calculate these sums with a very high accuracy. They must be used when no particular state provides a dominant contribution to the probability. In contrast to most of previous theoretical predictions the LOPT seems to hold up to very high intensity values. The experiments in neon and helium have confirmed the validity of the lowest-order theory.

The situation resulting from the presence of a resonance has been approached within a theory where all the higher-order contributions are exactly summed. The shifts of the resonance peaks show linear intensity dependences while the slopes of the curves representing the probability (or the ion number) as a function of the intensity, vary rapidly when one crosses the resonance. Such variations of the slope denote the presence of a resonance even when it is completely damped by intensity effects.

The time dependence of the probability is also strongly affected in resonant processes. It is no longer possible to define a time-independent transition rate in the whole range of photon energy. For 4-photon ionization of Cs we have shown the existence of two time variation regimes. In fact, the cubic time variation is characteristic of a single resonant process. In general, the probability must vary as the $(2Q - 1)$th power of time for small interaction times and/or near resonance, where Q is the number of principal poles. In the presence of a double resonance the probability would vary as the fifth power of time.

Owing to the difficulty of changing continuously the interaction time in experiments, the observation of these theoretical predictions has not been done. Nevertheless, a comparison regarding the time dependence of the resonance peaks has been possible. The agreement between the theoretical data and the experimental results obtained in the nanosecond and the picosecond range is excellent.

From the ensemble of results we have obtained, it appears that theory and experiment can be favorably compared even in the realistic situation where the interaction is instantaneously switched on and switched off (square pulse model). The predictions regarding photon correlation effects on the position and the amplitude of the resonance peaks are done within the same model. The agreement with the experimental results concerning the shifts and the amplitudes of the resonance peaks obtained with a coherent and a chaotic light justify the use of the zero-bandwidth model for the field.

The purpose of this discussion was to summarize the knowledge we have acquired about the laws governing nonresonant and resonant ionization of atoms. The next step in the research concerning multiphoton processes will be to utilize the techniques elaborated in atoms to treat the problems encountered in nonlinear interaction of light with complex media.

REFERENCES

Armstrong, L., Jr., Lambropoulos, P., and Rahman, N. K., (1976). *Phys. Rev. Lett.* **36**, 952.
Aymar, M., and Crance, M. (1979). *J. Phys. B* **12**, L667.
Bakos, J. S. (1974). *Adv. Electron. Electron. Phys.* **36**, 57.
Bebb, H. B. (1967). *Phys. Rev.* **153**, 23.
Bebb, H. B., and Gold, A. (1966). *Phys. Rev.* **143**, 1.
Beers, B. L., and Armstrong, L., Jr. (1975). *Phys. Rev. A* **12**, 2447.
Bethe, H. A., and Salpeter, E. E. (1957). "Quantum Mechanics of One and Two Electron Atoms." Academic Press, New York.
Cohen-Tannoudji, C. (1967). *Cargese Lect. Phys.* **2**, 347.
Cohen-Tannoudji, C., and Haroche, S. (1969). *J. Phys. (Orsay, Fr.)* **30**, 125, 153.
Crance, M., and Aymar, M. (1979). *J. Phys. B* **12**, 3665.
Dalgarno, A., and Lewis, J. T. (1955). *Proc. R. Soc. London, Ser. A* **233**, 70.
Debethune, J. L. (1972). *Nuovo Cimento B* **12**, 101.
Fedorov, M. V., and Kazakov, A. E. (1977). *Opt. Commun.* **22**, 42.
Georges, A. T., and Lambropoulos, P. (1980). *Adv. Electron. Electron Phys.* **54**, 191.
Glauber, R. J. (1963). *Phys. Rev.* **130**, 2529.
Glauber, R. J. (1965). "Quantum Optics and Electronics, Les Houches 1964" (C. Dewitt, A. Blandin, and C. Cohen-Tannoudji, eds.), pp. 63–185. Gordon and Breach, New York.
Goldberger, M. L., and Watson, K. M. (1964). "Collision Theory." Wiley, New York.
Gontier, Y., and Trahin, M. (1968). *Phys. Rev.* **172**, 83.
Gontier, Y., and Trahin, M. (1971). *Phys. Rev. A* **4**, 1896.
Gontier, Y., and Trahin, M. (1973). *Phys. Rev. A* **7**, 2069.
Gontier, Y., and Trahin, M. (1979a). *Phys. Rev. A* **19A**, 264.
Gontier, Y., and Trahin, M. (1979b). *J. Phys. B* **12**, 2123.
Gontier, Y., and Trahin, M. (1980). *J. Phys. B* **13**, 259.
Gontier, Y., Rahman, N. K., and Trahin, M. (1975). *Phys. Lett. A* **54A**, 341.
Gontier, Y., Rahman, N. K., and Trahin, M. (1976). *Phys. Rev. A* **14**, 2109.
Gontier, Y., Rahman, N. K., and Trahin, M. (1981a). *Lett. Nuovo Cimento* **32**, 348.
Gontier, Y., Rahman, N. K., and Trahin, M. (1981b). *Phys. Rev. A* **24**, 3102.
Göppert-Mayer, M. (1931). *Ann. Phys. (Leipzig)* **9**, 273.
Guccione-Gush, R., Gush, H. P., and Van Kranendonk, J. (1967). *Can. J. Phys.* **45**, 2513.
Karule, E. (1971). *J. Phys. B* **4**, L67.
Karule, E. (1974). "Atomic Processes." Report of the Latvian Academy of Sciences (USSR), pp. 5–24.
Klarsfeld, S. (1969). *Lett. Nuovo Cimento* **2**, 548.
Klarsfeld, S. (1970). *Lett. Nuovo Cimento* **3**, 395.
Lambropoulos, P. (1968). *Phys. Rev.* **168**, 1418.
Lambropoulos, P. (1976). *Adv. At. Mol. Phys.* **12**, 87.
Lambropoulos, P., Kikuchi, C., and Osborn, R. K. (1966). *Phys. Rev.* **144**, 1081.
Laplanche, G., Durrieu, A., Flank, Y., Jaouen, M., and Rachman, A. (1976). *J. Phys. B* **9**, 1263.
Lecompte, C., Mainfray, G., Manus, C., and Sanchez, F. (1974). *Phys. Rev. Lett.* **32**, 265.
Lecompte, C., Mainfray, G., Manus, C., and Sanchez, F. (1975). *Phys. Rev. A* **11**, 1009.
Lompré, L. A., Mainfray, G., Manus, C., and Thebault, J. (1978). *J. Phys. (Orsay, Fr,)* **39**, 610.

Lompré, L. A., Mainfray, G., Manus, C., and Marinier, J. P. (1981). *J. Phys. B* **14**, 4307.
McClean, W. A., and Swain, S. (1978). *J. Phys. B* **11**, L515.
Maquet, A. (1977). *Phys. Rev. A* **15A**, 1088.
Mollow, B. R. (1968). *Phys. Rev.* **175**, 1555.
Morellec, J., Normand, D., and Petite, G. (1976). *Phys. Rev. A* **14A**, 300.
Morellec, J., Normand, D., and Petite, G. (1979). *J. Phys. (Orsay, Fr.)* **40**, 172.
Morellec, J., Normand, D., and Petite, G. (1982). *Adv. At. Mol. Phys.* (to be published).
Mostowski, J. (1976). *Phys. Lett. A.* **56A**, 87.
Mower, L. (1966). *Phys. Rev.* **142**, 799.
Petite, G., Morellec, J., and Normand, D. (1979). *J. Phys. (Orsay, Fr.)* **40**, 115.
Schwartz, C. (1959). *Ann. Phys. (N. Y.)* **6**, 156.
Schwartz, C., and Tieman, J. J. (1959). *Ann. Phys. (N. Y.)* **6**, 178.
Shore, B. W., and Ackerhalt, J. (1977). *Phys. Rev. A* **15**, 1640.
Zernik, W. (1964). *Phys. Rev. A* **51**, 135.
Zernik, W., and Klopfenstein, R. W. (1965). *J. Math. Phys.* **6**, 262.
Zoller, P. (1979). *Phys. Rev. A* **19**, 1151.
Zon, B. A., Manakov, N. L., and Rapoport, L. P. (1972). *Sov. Phys. JETP (Engl. Transl.)* **34**, 515.
Zwanzig, R. (1960). *Lect. Theor. Phys.* **3**, 106.
Zwanzig, R. (1964). *Physica (Amsterdam)* **30**, 1109.

4

Calculation of Resonant Multiphoton Processes

MICHÈLE CRANCE

Laboratoire Aimé Cotton, CNRS II
Orsay, France

I. INTRODUCTION

A typical multiphoton ionization experiment consists of shining a strong, short pulse of light on a vapor. One then measures the number of either ions or electrons created, and possibly the angular distribution of the emitted electrons. The variation of these quantities is studied as a function of the characteristics of the light pulse: the frequency, intensity, linewidth, pulse length, polarization, and spatial distribution of intensity in the beam. When the frequency of the field is larger than $\omega_0 = E_0/\hbar$, where E_0 is the ionization energy of the ground state, the resulting ionization is referred to as photo-ionization. Photoionization may be interpreted as the absorption of one photon by an atom, the transition between the ground state and a continuum state being resonant for any frequency larger than ω_0. When the frequency of the field is smaller than ω_0, only the absorption of several photons can produce a resonant transition between the ground state and a continuum state. The minimum number of photons required for such a transition is n, the integer part of $\omega_0/\omega + 1$. This process is referred to as multiphoton ionization. For most frequencies, the n-photon transition from the ground state to continuum states is the resonant process of lowest possible order, which means that there is no atomic state with energy $\hbar\omega$, $2\hbar\omega, \ldots, (n-1)\hbar\omega$ above the ground state. The situation is then called nonresonant multi-photon ionization. A satisfactory description of the process is obtained by using perturbation theory at nth order. Any measured quantity factorizes as the product of several terms characterizing either the atom or the light pulse.

For certain frequencies, however, a Bohr frequency of the atom is reso-nant with the frequency of the field or a harmonic of it. For frequencies close to such resonances, a perturbative description is no longer possible. Such a situation is referred to as resonant multiphoton ionization. An unusual case of resonant multiphoton ionization occurs when an autoionizing state is involved. This case presents some particular aspects which justify a special treatment; Chapter 8 by Lambropoulos and Zoller is especially devoted to this case, so we shall not consider it further. There have been a few works treating the very strong field limit, which is when the electron light inter-action is much larger than the electron interaction (Austin, 1979; Geltman, 1977; Goldberg and Shore, 1978; Brandi and Davidovitch, 1979; Faisal, 1973; Faisal and Moloney, 1981; Geltman and Teague, 1974; Reiss, 1980). We shall not discuss these works since the formalism is very different from the one considered here, and none of these works has reached the point where any comparison with experiment is possible.

Theoretical studies that attempt to quantitatively describe the resonant processes proceed in several steps. First, one defines a model to describe the process, which allows one to determine the relevant parameters with which to characterize the atom and the light. These parameters then must be calculated, after which one has to study the dynamics of the process in order to obtain the number of ions created or to evaluate any other measured quantity.

A large number of papers have been devoted to the study of some of these steps, especially in the last few years, and, even though many questions remain unanswered, the most striking features of resonant multiphoton ionization are now understood. Various methods have been proposed which give equivalent results. The purpose of this paper is to indicate the scope of the tools available for the interpretation of resonant multiphoton ionization experiments.

The subject has already been reviewed several times (Bakos, 1974; Delone, 1975; Lambropoulos, 1976, 1980; Georges and Lambropoulous, 1980; Mainfray and Manus, 1980). The understanding of multiphoton ionization has grown a little since these reviews appeared, essentially owing to the advent of experiments in which absolute measurements have been performed (Lompré *et al.*, 1980; Normand and Morellec, 1981). Such accurate experimental results produce extremely valuable information that has stimulated new atomic structure calculations. The interference minimum observed in 2-photon ionization of the cesium ground state is a good example of this situation. In this chapter we will present the framework of these new calculations. We will recall the basic principles of these methods, even though this presentation may sometimes appear elementary. We felt this method necessary in order for the presentation to be self-contained, and, furthermore, to give an additional point of view that may help the reader to find his way in the maze of resonant multiphoton ionization.

First, we recall briefly how nonresonant processes are described (Section II). This provides the opportunity to explain what quantities are measured in an experiment and to point out what assumptions of nth-order perturbation treatment break down when a resonance occurs. This introduction will serve as a guide to the study of resonant processes. A realistic description of the resonant process requires sophisticated models for both the atom and the light field. To split up the difficulty, we shall at first use an oversimplified model for the light field—a monochromatic field with rectangular pulse shape—to define a simple but realistic model for the atom (Section III). With this model, we define an effective Hamiltonian which is used in Section IV to study the dynamics of the process. We then discuss how a more sophisticated description of the light pulse may be introduced, and consider how the previous results are modified when the spatial distribution of intensity is taken into account (Section V). In order to make a precise comparison with experiments, it is necessary to calculate accurately the atomic parameters defined in Sections III and IV. In Section VI we present the methods developed to perform such calculations. In Section VII we consider the case in which electron yield is analyzed.

II. NONRESONANT PROCESSES

In this Section we shall recall briefly how nonresonant multiphoton ionization is interpreted. This will give us the opportunity to review the successive steps of a complete calculation of multiphoton ionization. Moreover, we shall discuss the validity of the basic approximations used to calculate nonresonant processes. In this perspective, resonant multiphoton ionization is a particular case in which these basic assumptions break down.

The ionization probability $P(I)$ per unit time calculated by using the perturbation theory at nth order is given by

$$P(I) = \sigma^{(n)} I^n \tag{2.1}$$

(see Section III.4), where I is the field intensity; $\sigma^{(n)}$, the generalized cross section, is given by

$$\sigma^{(n)} = \frac{2}{\hbar}\left(\frac{2}{\varepsilon_0 c}\right)^n$$

$$\times \sum_{\varepsilon}\left(\sum_{i,j,\ldots,k}\frac{\langle g|d|i\rangle\langle i|d|j\rangle\cdots\langle k|d|\varepsilon\rangle}{(E_g+\hbar\omega-E_i)(E_g+2\hbar\omega-E_j)\cdots(E_g+(n-1)\hbar\omega-E_k)}\right)^2 \tag{2.2}$$

where $|g\rangle$ is the ground state, or more generally the initial state; d is the atomic dipole operator; $|i\rangle, |j\rangle, \ldots$ are all the atomic states; $|\varepsilon\rangle$ are the continuum states with energy $E_g + n\hbar\omega$, E_g being the energy of $|g\rangle$.

The intensity incident on an atom is a function of the time t and the position $\mathbf{r}$ where the atom is located. (We neglect the motion of an atom during the interaction time.) It may be written as $I = I_M f(t) g(\mathbf{r})$. I_M is the maximum intensity with respect to time and space, such that the maximum values of both $f(t)$ and $g(\mathbf{r})$ are 1. The probability to be ionized by a light pulse for an atom at $\mathbf{r}$ is

$$1 - \exp\left(-\int_{-\infty}^{+\infty} \sigma^{(n)} I_M^n g^n(\mathbf{r}) f^n(t)\, dt\right) \tag{2.3}$$

which reduces to

$$\int_{-\infty}^{+\infty} \sigma^{(n)} I_M^n\, g^n(\mathbf{r}) f^n(t)\, dt \tag{2.4}$$

when the ionization probability is low enough, that is, when the process is far from saturation. When Eq. (2.1) is valid, the time dependence of the light pulse can be characterized, in so far as ionization is concerned, by the parameter

$$\tau_n = \int_{-\infty}^{+\infty} f^n(t)\, dt \tag{2.5}$$

The number of ions created by a light pulse is

$$N_E = N_0 \int_v \{1 - \exp[-\sigma^{(n)} I_M^n\, g^n(\mathbf{r})\tau_n]\}\, d\mathbf{r} \tag{2.6a}$$

which reduces to

$$N_E = N_0 \sigma^{(n)} I_M^n \tau_n \int_v g^n(\mathbf{r})\, d\mathbf{r} \tag{2.6b}$$

when Eq. (2.4) is valid. In Eqs. (2.6) N_0 is the number of atoms per unit volume and the integral runs over the volume of the vapor. Before saturation is reached, the spatial distribution is characterized by the parameter

$$V_n = \int_v g^n(\mathbf{r})\, d\mathbf{r} \tag{2.7}$$

so that, finally,

$$N_E = N_0 \sigma^{(n)} I_M^n \tau_n V_n \tag{2.8}$$

in the limit of weak intensities.

It is difficult to define a weak field. If we define a weak field as a field for which Eq. (2.6a) is valid, then the critical values vary by several orders of magnitude, depending on the number of photons absorbed and on the atom and the frequency. This is easily understood from the structure of Eq. (2.2), where the energy denominators may change by $\sigma^{(n)}$ over a large range. The basic approximation of a perturbation treatment at minimum nonvanishing order may fail in two ways: either perturbative treatment is valid but higher orders must be introduced, or perturbation treatment is not appropriate and another approach must be developed.

Even if it is still meaningful to interpret the process by calculating an ionization probability per unit time, keeping only the lowest-order expression for $\sigma^{(n)}$ may be insufficient. However, calculating $\sigma^{(n)}$ in lowest order is already complicated, and if higher orders of perturbations have to be introduced, the complexity rapidly increases! In most cases it is hopeless. However, the problem can be handled if only some types of terms in the perturbation series dominate. We may envision re-summing these terms in a closed form. This situation occurs when the large magnitude of $\sigma^{(n)}$ is related to the presence of one (or a few) small energy denominators. If the denominator $E_g + p\hbar\omega - E_r$ is small the p-photon transition $g-r$ is nearly resonant.

If we now reconsider the case where a p-photon transition is nearly resonant, we expect to observe the features of a two-level system, i.e., Rabi oscillations. Qualitatively, the largest contribution to the ionization probability of the atom in the resonant case is the product of the population of the resonant state by its ionization probability. Because of Rabi oscillations of the populations, we expect the ionization probability of the atom to be an oscillating function of time. However, in some cases the oscillation may not be the dominant feature, as when for example, the Rabi period is much larger than the interaction time, or conversely, the Rabi period is much smaller than the interaction time and many oscillations occur before ionization is saturated. In the latter case we may replace the oscillating ionization probability by its average over a few Rabi periods in interpreting the experimental data.

III. DEFINITION OF AN EFFECTIVE HAMILTONIAN

In order to describe the dynamics of resonant multiphoton ionization, we need to define a model for the atom and a model for the field. In this Section we use a simple model for the excitation field—a constant monochromatic field which is switched on at $t = 0$ and switched off at $t = T$—to define a realistic model for the atom. Various formalisms may be used, which fundamentally lead to the same final result. After presenting one of them, we shall briefly outline alternative approaches.

We first analyze the atom-plus-field system being described in the dressed atom picture. By using the resolvent operator formalism (Goldberger and Watson, 1964; Cohen-Tannoudji, 1967; Cohen-Tannoudji et al. 1969a,b; Mower, 1966), we form a model for the atom and define the relevant atomic quantities to be introduced.

A. Analysis of the Process

The Hamiltonian for the atom-plus-field system is

$$H = H_0 + V \qquad (3.1)$$

$$H_0 = H_A + H_F \qquad (3.2)$$

H_A is the atomic Hamiltonian and H_F is the field Hamiltonian (Messiah, 1965)

$$H_F = \hbar\omega a^+ a \qquad (3.3)$$

where ω is the field frequency and V is the atom field interaction. For the sake of simplicity we shall only consider the dipole interaction

$$V = \sqrt{2\hbar\omega/\varepsilon_0 v}\, d(a + a^+) \qquad (3.4)$$

where d is the atomic dipole operator; a and a^+ are the annihilation and creation operator of the field, respectively; and v is the quantization volume. Higher terms of multipole expansion of V could be easily introduced in the formalism we describe; this would introduce some new atomic parameters but would not change the structure of the equation of motion that we have derived. Moreover, such contributions are expected to be important only in the close vicinity of a resonance forbidden by the dipole interaction (Jaouen et al., 1980). They might also be important in strongly nonresonant situations, as when the dipole approximation predicts a deep minimum (Lambropoulos et al., 1975; Flank et al., 1976).

The eigenfunctions of H_0 are $|i\rangle|N\rangle$, where $|i\rangle$ is an atomic state and $|N\rangle$ is a field state in number representation. $|g\rangle$ is the initial state. The eigenenergy of state $|i\rangle|N\rangle$ is $E_i + N\hbar\omega$. The fact that n photons are required to ionize $|g\rangle$ means that the state $|g\rangle|N\rangle$ is degenerate with the states $|\varepsilon\rangle|N - n\rangle$, $|\varepsilon\rangle$ being a state of the atomic continuum with energy $E_\varepsilon \simeq E_g + n\hbar\omega$. In the same way, any state $|i\rangle|N\rangle$ associated with a discrete state $|i\rangle$ is degenerate with some continuum state $|\varepsilon\rangle|N'\rangle$. As has been pointed out by Armstrong et al. (1975), this situation is similar to autoionization in atoms. Each discrete eigenstate of H_0 that is coupled to an ionization continuum causes an unstable state, whose relaxation probability is to be interpreted as the ionization probability of the atomic state. These states, called pseudoautoionizing states by Armstrong et al., might then be treated by using the formalism developed by Fano (1961) for atomic autoionized states.

A resonance occurs when a state $|r\rangle|N - p_r\rangle$ is quasidegenerate with $|g\rangle|N\rangle$, i.e., when

$$E_g \simeq E_r - p_r\hbar\omega \tag{3.5}$$

the p_r-photon transition $g-r$ is resonant (or nearly resonant). The evolution of the atom-plus-field system is described by some wave function $|\psi(t)\rangle$. When $t = 0$, $|\psi(t)\rangle = |g\rangle|N_0\rangle$. After the interaction has been initiated and the states $|g\rangle|N_0\rangle$ and $|r\rangle|N_0 - p_r\rangle$ are coupled, the weights of these states in $|\psi(t)\rangle$ may have comparable magnitude. More generally, we expect that the weight in $|\psi(t)\rangle$ of a state $|i\rangle|N_0 - p_i\rangle$ may be large when the condition

$$E_g \simeq E_i - p_i\hbar\omega \tag{3.6}$$

is satisfied. On the other hand, we expect the projection of $|\psi(t)\rangle$ on states $|\alpha\rangle|N_0 - p_\alpha\rangle$ to be small when

$$|E_g + p_\alpha\hbar\omega - E_\alpha| \tag{3.7}$$

is large. The projection of $|\psi(t)\rangle$ on a continuum state $|\varepsilon\rangle|N\rangle$ represents the probability amplitude for the emission of an electron in state $|\varepsilon\rangle$. The probability for the atom to be ionized at time t is then

$$\mathcal{N}_i(t) = \sum_N \int d\varepsilon \, |\langle\psi(t)|\varepsilon\rangle|N\rangle|^2 \qquad (3.8)$$

The summation runs over any N and any atomic continuum state. By using the closure relation, we rewrite Eq. (3.8) as

$$\mathcal{N}_i(t) = 1 - \sum_N \sum_i |\langle\psi(t)|i\rangle|N\rangle|^2 \qquad (3.9)$$

The summation runs over any N and any discrete atomic state. A perturbative treatment at lowest order would consist of keeping only the leading term of $1 - |\langle\psi(t)|g\rangle|N_0\rangle|^2$ in Eq. (3.9). Such an approximation is fully justified in nonresonant situations since no discrete state satisfies condition (3.6). We discussed above the relative orders of magnitude of the remaining terms in a resonant situation. Following this analysis, it is consistent to keep all the terms $\langle\psi(t)|i\rangle|N_0 - p_i\rangle$ in Eq. (3.9) corresponding to the nearly resonant states in Eq. (3.6). These states and $|g\rangle|N_0\rangle$ span a space $\mathscr{E}_R$. The projection operator on $\mathscr{E}_R$ is

$$\mathbb{P} = |g\rangle|N_0\rangle\langle g|\langle N_0| + \sum_i |i\rangle|N_0 - p_i\rangle\langle i|\langle N_0 - p_i| \qquad (3.10)$$

where i and p_i satisfy Eq. (3.6). The projection operator on the complementary space is

$$\mathbb{Q} = \mathbb{I} - \mathbb{P} \qquad (3.11)$$

Our approximation consists in defining the probability for an atom to be ionized at t by

$$\mathcal{N}_i(t) = 1 - \langle\psi(t)|\mathbb{P}|\psi(t)\rangle \qquad (3.12)$$

and thus we are looking for the evolution of $\mathbb{P}|\psi(t)\rangle$.

We shall use the resolvent operator formalism to derive an effective Hamiltonian acting on space $\mathscr{E}_R$. This formalism is described in detail by Goldberger and Watson (1964); we recall briefly the basic formulas to be used in our notation.

B. Definition of an Effective Hamiltonian

Introducing the evolution operator $U(t, t')$, the wave function may be rewritten as

$$|\psi(t)\rangle = U(t, 0)|\psi(0)\rangle \qquad (3.13)$$

The resolvent operator $G(z)$ is defined by

$$G(z) = (z - H)^{-1} \qquad (3.14)$$

For $t > 0$, it is related to the evolution operator by

$$U(t, 0) = \lim_{y \to 0_+} \left\{ -\frac{1}{2i\pi} \int_{-\infty}^{+\infty} G(x + iy)e^{-i(x+iy)t} \, dx \right\} \qquad (3.15)$$

where x and y are real and the limit goes to zero when $y > 0$.

To study $\mathbb{P}|\psi(t)\rangle$, we need to evaluate $\mathbb{P}U(t, 0)\mathbb{P}$, and thus $\mathbb{P}G\mathbb{P}$. The resolvent operator associated with H_0 is

$$G_0 = (z - H_0)^{-1} \qquad (3.16)$$

G and G_0 are related by

$$G = G_0 + G_0 V G \qquad (3.17)$$

As projection operators $\mathbb{P}$ and $\mathbb{Q}$ satisfy the relations

$$\mathbb{P}^2 = \mathbb{P}, \qquad \mathbb{Q}^2 = \mathbb{Q}, \qquad \mathbb{P}\mathbb{Q} = \mathbb{Q}\mathbb{P} = 0 \qquad (3.18)$$

Moreover, H_0 and G_0 are diagonal in the basis $|i\rangle|N\rangle$ and thus

$$\mathbb{P}G_0\mathbb{Q} = \mathbb{Q}G_0\mathbb{P} = 0 \qquad (3.19)$$

Starting from

$$\mathbb{P}(z - H_0)G\mathbb{P} = \mathbb{P} + \mathbb{P}VG\mathbb{P} \qquad (3.20)$$

and using Eqs. (3.11), (3.18), and (3.20), we derive the successive equations

$$\mathbb{P}(z - H_0 - V)\mathbb{P}G\mathbb{P} = \mathbb{P} + \mathbb{P}V\mathbb{Q}G\mathbb{P} \qquad (3.21)$$

$$\mathbb{P}(z - H_0 - V - V\mathbb{Q}G_0\mathbb{Q}V)\mathbb{P}G\mathbb{P} = \mathbb{P} + \mathbb{P}V\mathbb{Q}G_0\mathbb{Q}V\mathbb{Q}G\mathbb{P} \qquad (3.22)$$

This can be rewritten as

$$\mathbb{P}[z - H_0 - R(z)]\mathbb{P}G\mathbb{P} = \mathbb{P} \qquad (3.23)$$

where

$$R(z) = V + \sum_{q=1}^{\infty} V(\mathbb{Q}G_0\mathbb{Q}V)^q \qquad (3.24)$$

In space $\mathscr{E}_R$, $\mathbb{P}$ is the unit operator. The restriction of G to the space $\mathscr{E}_R$ is thus

$$\mathbb{P}G\mathbb{P} = \mathbb{P}(z - H_0 - R(z))^{-1}\mathbb{P} \qquad (3.25)$$

This expression has been obtained by Gontier *et al.* (1976) in a formalism based on a diagrammic re-summation of the perturbation expansion. We expect that the largest contribution to $\mathbb{P}U(t, 0)\mathbb{P}$ comes from poles in $\mathbb{P}G\mathbb{P}$ close to

$$E_0 = E_g + N_0\hbar\omega \qquad (3.26)$$

Due to the structure of $R(z)$, all the poles of $R(z)$ are far from E_0. Moreover, we expect $R(z)$ to be a slowly varying function of z in the vicinity of E_0. We thus can replace z by E_0 in $R(z)$ in lowest order. Now $\mathbb{P}G\mathbb{P}$ has the structure of a resolvent operator in $\mathscr{E}_R$, corresponding to the Hamiltonian

$$H_0 + R(E_0) \qquad (3.27)$$

This is essentially the effective Hamiltonian for which we were looking. However, we must keep in mind that the replacement $z \to E_0$ is only approximate; we might as well have replaced z by any of the resonant energies $E_i + (N_0 - p_i)\hbar\omega$. Because $R(E_0)$ is an infinite series, we shall keep only the leading terms to build an effective Hamiltonian. This calls for a careful examination of the structure of $R(E_0)$.

C. Structure of the Effective Hamiltonian

Apart from V, each term of $\mathbb{P}R(E_0)\mathbb{P}$ contains $\mathbb{Q}G_0\mathbb{Q}$, that is, a sum over all the discrete nonresonant states $|\alpha\rangle|N_0 - p_\alpha\rangle$ and integrals over any continuum $|\varepsilon\rangle|N\rangle$. When N is equal to $N_0 - n$ or smaller, $\mathbb{Q}G_0\mathbb{Q}$ exhibits a divergence for

$$E_\varepsilon = E_g + (N_0 - n - N)\hbar\omega \qquad (3.28)$$

To understand how this divergence has to be treated, we have to go back to Eq. (3.15). Then z is to be replaced by $x + iy$ (x, y, real; y, positive), and the limit when y goes to zero is to be taken. We thus have to replace the divergent term $(E_\varepsilon + N\hbar\omega - E_0)^{-1}$ in $\mathbb{Q}G_0(E_0)\mathbb{Q}$ by

$$\mathscr{P}(E_\varepsilon + N\hbar\omega - E_0)^{-1} - i\pi\delta(E_\varepsilon + N\hbar\omega - E_0) \qquad (3.29)$$

(Heitler, 1954), where $\mathscr{P}$ implies that the Cauchy principle value will be taken when integrating over ε. As a consequence, any matrix element of $\mathbb{P}R(E_0)\mathbb{P}$ is the sum of an infinite series of real terms and an infinite series of imaginary terms. In building the effective Hamiltonian H_{eff}, for each matrix element of $\mathbb{P}R(E_0)\mathbb{P}$ we shall keep the leading real term and the leading imaginary term, which are not necessarily of the same order in V. Let us call $\hat{R}(E_0)$ the resulting matrix.

The real part of any diagonal element of $\hat{R}(E_0)$ is of second order in V; this is the quadratic term of the light shift, in which the contribution of resonant states has been omitted. For example,

$$\text{Re}\langle g|\langle N_0|\hat{R}(E_0)|g\rangle|N_0\rangle$$

$$= \sum_u \frac{|\langle g|\langle N_0|V|u\rangle|N_0 + 1\rangle|^2}{E_g - \hbar\omega - E_u} + \sum_{u \notin \mathscr{E}_R} \frac{|\langle g|\langle N_0|V|u\rangle|N_0 - 1\rangle|^2}{E_g + \hbar\omega - E_u} \qquad (3.30)$$

in the first term u stands for any discrete or continuum atomic state; in the second term the quasi-resonant states belonging to the photon occupation number $|N_0 - 1\rangle$ have been excluded if such states exist.

The order of an imaginary diagonal term in V for a particular state depends on how many photons are required to ionize the atom from this state. It is $(n - p_i)$ for $|i\rangle|N_0 - p_i\rangle$. Then

$$\text{Im}\langle i|\langle N_0 - p_i|\hat{R}(E_0)|i\rangle|N_0 - p_i\rangle = \sum_\varepsilon [W_{i, N_0 - p_i; \varepsilon, N_0 - n}]^2 \qquad (3.31)$$

The sum runs over all the resonant continuum states $|\varepsilon\rangle|N_0 - n\rangle$

$W_{i, N_0 - p_i; \varepsilon, N_0 - n}$

$$= \sum_{\alpha, \beta, \ldots, \gamma} \frac{\langle i|\langle N_0 - p_i|V|\alpha\rangle|N_0 - p_i - 1\rangle\langle\alpha|\langle N_0 - p_i - 1|V|\beta\rangle|N_0 - p_i - 2\rangle}{(E_g - (p_i + 1)\hbar\omega - E_\alpha)(E_g - (p_i + 2)\hbar\omega - E_\beta)} $$
$$\frac{\cdots \langle\gamma|\langle N_0 - n + 1|V|\varepsilon\rangle|N_0 - n\rangle}{\cdots (E_g - (n - 1)\hbar\omega - E_\gamma)}$$

$$(3.32)$$

The $(n - p_i - 1)$ states $\alpha, \beta, \ldots$ appear in the summation. For each multiplicity the states belonging to $\mathscr{E}_R$ are excluded. The real part of a nondiagonal term is the probability amplitude of the lowest possible order multiphoton transition connecting the corresponding atomic states, in which the contribution of quantum paths involving other resonant states has been omitted. The matrix element between $|g\rangle|N_0\rangle$ and $|i\rangle|N_0 - p_i\rangle$ is of order p_i in V. However, the order of the transition between two resonant states $|i\rangle|N_0 - p_i\rangle$ and $|j\rangle|N_0 - p_j\rangle$ is not always simply related to p_i and p_j because of the selection rules for the dipole interaction.

The imaginary part of a nondiagonal matrix element between two states $|i\rangle|N_0 - p_i\rangle$, $|j\rangle|N_0 - p_j\rangle$ or $|g\rangle|N_0\rangle$ is

$$\sum_\varepsilon W_{i, N_0 - p_i; \varepsilon, N_0 - n} W_{j, N_0 - p_j; \varepsilon, N_0 - n} \qquad (3.33)$$

The sum runs over all the resonant continuum states $|\varepsilon\rangle|N_0 - n\rangle$ with energy given by Eq. (3.28). Let us remark that in keeping only the leading imaginary term, we systematically neglect the ionization towards continua higher than the lowest accessible one. These are reached by absorption of $n + 1, n + 2, \ldots$ photons.

The terms of H_{eff} we have discussed have a well-defined order in V and thus they factorize in a field term and quantity depending only on the atom structure and the field frequency. The field term is a product such as $[N_0(N_0 - 1) \cdots]^{1/2}$. Owing to the large magnitude of N_0 in the range where multiphoton ionization is observed, the field term can be replaced by the corresponding power of $N_0^{1/2}$ which is related to the field intensity by

$$I = N_0 \hbar \omega c / v \tag{3.34}$$

There is no need to know the atomic parameters in order to discuss the dynamics of resonant multiphoton ionization, so we shall defer discussion of methods for calculating atomic quantities to Section VI.

The problem remaining is how to choose the states to be introduced in $\mathscr{E}_R$, that is to specify what Eq. (3.6) means. For a given frequency or frequency range it is easy to recognize which states must be included in $\mathscr{E}_R$ by examining the energy spectrum near energies $E_0 + \hbar \omega$, $E_0 + 2\hbar \omega$, $\cdots$. One must keep in mind, however, that a nonresonant state in weak field may become resonant in strong field because of the ac-Stark shift of the transition [diagonal real elements of $\hat{R}(E_0)$]. More precisely, as can be seen from the study of the dynamics of the process given below, a state has to be introduced in $\mathscr{E}_R$ when either its shift or ionization probability is of the order of the detuning from exact resonance. Apart from this criterion, a good way to limit $\mathscr{E}_R$ is to introduce more and more states until the final result becomes stable with respect to the size of the basis. Fortunately, when the frequency is varied in a narrow range that includes a resonant state, it is generally not necessary to introduce any further resonant state.

The effective Hamiltonian that we have defined depends on the field frequency through E_0. This dependence may be neglected when the frequency is varied in a narrow range. However, if we intend to study a large range of frequency, it is necessary to take into account the variation of H_{eff} with the frequency. Actually, the structure of H_{eff} is such that the frequency dependence of matrix element H_{eff} is smooth, and it is possible to obtain a reliable effective Hamiltonian in a large range of frequency by interpolation based on a few frequencies (Crance and Aymar, 1980).

D. Nonresonant Ionization Probability

Finally, we can easily deduce the ionization probability of the initial state as calculated at nth order of perturbation from the calculation of H_{eff}. When no resonance occurs, $\mathscr{E}_R$ reduces to $|g\rangle|N_0\rangle$, and Eq. (3.28) becomes

$$\sum_\varepsilon \left[W_{g,N_0;\varepsilon, N_0-n} \right]^2 \tag{3.35}$$

with

$W_{g,N_0;\varepsilon, N_0-n}$

$$= \sum_{\alpha,\beta,\ldots,\gamma} \frac{\langle g|\langle N_0|V|\alpha\rangle|N_0-1\rangle \langle \alpha|\langle N_0-1|V|\beta\rangle|N_0-2\rangle}{(E_g + \hbar\omega - E_\alpha)(E_g + 2\hbar\omega - E_\beta)} \cdots \langle \gamma|\langle N_0-n+1|V|\varepsilon\rangle|N_0-n\rangle}{\cdots [E_g + (n-1)\hbar\omega - E_\gamma]} \tag{3.36}$$

The summation now includes every state $|\alpha\rangle|N\rangle$. Equation (3.35) factorizes into a part independent of field [$\sigma^{(n)}$, Eq. (2.2)] and the nth power of the field intensity.

IV. DYNAMICS OF RESONANT MULTIPHOTON IONIZATION

In Section III we have purposely considered the most general case where several resonances may occur. However, it often turns out that only one resonance has to be considered in order to explain experimental results. Moreover, the most important general features of resonant multiphoton ionization may be explained by using a simple model, where only one resonance is taken into account and one continuum may be reached. However, although most important features of resonant multiphoton ionization may be derived from the study of the wave function $\mathbb{P}|\psi(t)\rangle$ [Eq. (3.9)], some problems, such as the effect of spontaneous emission or the statistics of the field, require the study of a density matrix. In Section IV.A we shall discuss the evolution of $\mathbb{P}|\psi(t)\rangle$ and in Section IV.B we introduce a density matrix formalism to discuss the remaining problems.

A. Evolution of the Wave Function

We first consider the case of a monochromatic light pulse with a rectangular time dependence (Beers and Armstrong, 1975; Feneuille and Armstrong, 1976). The ground state is $|g\rangle|N_0\rangle$, the resonant state is $|r\rangle|N_0 - p\rangle$, $P\psi(t)$ is given by

$$\mathbb{P}|\psi(t)\rangle = a_g|g\rangle|N_0\rangle + a_r|r\rangle|N_0 - p\rangle \tag{4.1}$$

The relevant matrix elements may be written

$$\langle g|\langle N_0|H_{\text{eff}}|g\rangle|N_0\rangle = \hbar(\Delta_g - iJ^2)$$
$$\langle r|\langle N_0 - p|H_{\text{eff}}|r\rangle|N_0 - p\rangle = \hbar(\Delta_r - iL^2)$$
$$\langle g|\langle N_0|H_{\text{eff}}|r\rangle|N_0 - p\rangle = \hbar(K - iJL) \tag{4.2}$$
$$\langle r|\langle N_0 - p|H_{\text{eff}}|g\rangle|N_0\rangle = \hbar(K - iJL)$$

where $\hbar\delta = E_r - E_g - p\hbar\omega$ defines the detuning from zero field resonance; Δ_g and Δ_r are the nonresonant parts of the light shift of $|g\rangle$ and $|r\rangle$, which vary as the field intensity I; K varies as $I^{p/2}$; J varies as $I^{n/2}$; L varies as $I^{(n-p)/2}$. (See Fig. 1.) The origin of energies is taken as E_0 [Eq. (3.26)]. The equation of motion for a_g and a_r may be written as

$$i\dot{a}_g = (\Delta_g - iJ^2)a_g + (K - iJL)a_r$$
$$i\dot{a}_r = (\Delta_r + \delta - iL^2)a_r + (K - iJL)a_g \tag{4.3}$$

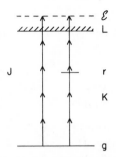

Fig. 1. Level scheme of the system governed by Eqs. (4.3) and (4.4).

$$K = \langle g | \langle N_0 | \hat{R}(E_0) | r \rangle | N_0 - p \rangle, \qquad L = \langle r | \langle N_0 - p | \hat{R}(E_0) | \varepsilon \rangle | N_0 - n \rangle,$$

$$J = \langle g | \langle N_0 | \hat{R}(E_0) | \varepsilon \rangle | N_0 - n \rangle$$

Each $\hat{R}(E_0)$ matrix element is the leading term of the corresponding $R(E_0)$ matrix element [Eq. (3.24)].

At $t = 0$ the system is in state $|g\rangle |N_0\rangle$ and thus $a_g(0) = 1$ and $a_r(0) = 0$. The probability for an atom to be ionized at time t is

$$\mathcal{N}_1(t) = 1 - |a_g|^2 - |a_r|^2 \tag{4.4}$$

A time-dependent ionization probability may be defined by

$$d\mathcal{N}_1(t)/dt = 2|Ja_g + La_r|^2 \tag{4.5}$$

The solution has been calculated by Beers and Armstrong (1975):

$$\mathcal{N}_1(t) = 1 - \frac{1}{|\lambda_+ - \lambda_-|^2} \left[|(\Delta_g - iJ^2 - \lambda_-)e^{-i\lambda_+ t} - (\Delta_g - iJ^2 - \lambda_+)e^{-i\lambda_- t}|^2 \right.$$
$$\left. + (K^2 + J^2 L^2)|e^{-i\lambda_+ t} - e^{-i\lambda_- t}|^2 \right] \tag{4.6a}$$

where

$$\lambda_{\pm} = \frac{\Delta_g + \Delta_r + \delta - iL^2 - iJ^2}{2}$$
$$\pm \left[\left(\frac{\Delta_g - \Delta_r - \delta + iL^2 - iJ^2}{2} \right)^2 + (K - iJL)^2 \right]^{1/2} \tag{4.6b}$$

This is the general solution to Eq. (4.3), which is valid under any conditions. In some cases, however, we may be able to reduce Eq. (4.5) to a simpler expression. In particular, we shall try to find when a time-independent ionization probability may be defined. The states $|g\rangle |N_0\rangle$ and $|r\rangle |N_0 - p\rangle$ give rise to pseudoautoionized states with energy Re $\lambda_{\pm} \hbar$ and ionization probabilities $2|\text{Im } \lambda_{\pm}|$. Let us first consider the oversimplified case where $J = 0$

and the field is exactly on resonance, i.e.,

$$\Delta_g = \Delta_r + \delta \qquad (4.7)$$

then

$$\lambda_\pm = -\tfrac{1}{2}iL^2 \pm (K^2 - \tfrac{1}{4}L^4)^{1/2} \qquad (4.8)$$

We thus see that the crucial point is the relative magnitude of K and $\tfrac{1}{2}L^2$. If K is much smaller than $\tfrac{1}{2}L^2$, one eigenstate is almost stable (its ionization probability is $2K^2/L^2$), the other one is unstable. If K is much larger than $\tfrac{1}{2}L^2$, the ionization probability from both eigenstates are equal and Rabi oscillations will dominate the evolution of $\mathscr{N}_1(t)$. When either $K \ll \tfrac{1}{2}L^2$ or $K \gg \tfrac{1}{2}L^2$, Eqs. (4.5) and (4.6) reduce to a simple form, which we shall discuss in detail below. The situation is also simpler when the interaction time is much smaller than any of the characteristic times: K^{-1}, L^{-2}.

1. Resonant State Weakly Coupled to the Initial State

$$K \ll \tfrac{1}{2}L^2 \qquad (4.9)$$

Owing to the number of photons involved, it is reasonable to assume that J^2 is much smaller than L^2. The imaginary parts of $\lambda_\pm$ [Eq. (4.6)] are then approximately given by, respectively,

$$\Gamma_+ = L^2 \qquad \text{and} \qquad \Gamma_- = \frac{[(\delta + \Delta_r - \Delta_g)J - KL]^2}{(\delta + \Delta_r - \Delta_g)^2 + L^4} \qquad (4.10)$$

Because of condition (4.9), Γ_+ is much larger than Γ_-. Then, for interaction times larger than Γ_-^{-1}, both exponentials vanish in Eq. (4.6), meaning that ionization is saturated: $\mathscr{N}_1(t) = 1$. For interaction times larger than Γ_+^{-1} but smaller than Γ_-^{-1}, $\exp(-i\lambda_+ t)$ is very small and the number of ions created reduces to

$$\mathscr{N}_1(t) = 1 - \exp(-2\Gamma_- t) \qquad (4.11)$$

Thus in the range of interaction time defined by

$$\Gamma_+^{-1} \ll t \ll \Gamma_-^{-1} \qquad (4.12)$$

a time-independent ionization probability may be defined, i.e.,

$$d\mathscr{N}_1(t)/dt = 2\Gamma_- \qquad (4.13)$$

This situation typically happens when p is larger than $(n - p)$. Then Γ_+ and Γ_- may differ by several orders of magnitude and the range defined by Eq. (4.12) is very large.

The *profile* obtained by recording the number of ions created as a function of the frequency is described by

$$\frac{d\mathcal{N}_1}{dt} = 2\frac{[(\delta + \Delta_r - \Delta_g)J - KL]^2}{(\delta + \Delta_r - \Delta_g)^2 + L^4} \qquad (4.14)$$

As long as the frequency dependence of H_{eff} may be neglected, Eq. (4.14) describes a Fano profile, as is obtained in the case of atomic autoionizing resonances. The Fano parameter here is

$$q = -K/JL \qquad (4.15)$$

The characteristic patterns of a Fano profile are a resonant peak and a zero minimum superimposed on a constant background. If the minimum and the maximum occur for frequencies too far from each other for the frequency dependence of H_{eff} to be neglibible, the profile is not, strictly speaking, a Fano profile (Dixit and Lambropoulos, 1979). We shall not discuss this point in detail since we are essentially interested in the resonant part of the profile. Equation (4.14) describes the superposition of a Lorentzian and a dispersion shape on a constant background. The Lorentzian is centered on

$$\delta = \Delta_g - \Delta_r \qquad (4.16)$$

and its width is $2L^2$. The Lorentzian shape represents the contribution of the resonant quantum paths leading to ionization. The background represents the contribution of nonresonant paths. Both types of quantum paths lead to the same ionization continuum and the probability amplitudes associated to these paths interfere. The corresponding contribution is the dispersive part of Eq. (4.14). Strictly speaking Eq. (4.14) always describes an asymmetric shape; however, the asymmetry may be observed only if the contrast of the resonance (measured by q^2) is weak enough. Note that q^2 is proportional to $I^{2(p-n)}$. For weak intensity the profile is almost Lorentzian, the maximum value varies as I^{2p-n}; when the intensity is increased, the contrast of the resonance decreases and the asymmetry becomes more noticeable until the resonance disappears when q^2 becomes smaller than 1.

The location of the resonance is given by Eq. (4.16). As Δ_r and Δ_g are linear functions of I, the location of the resonance varies when the intensity is varied.

Effective Order of Nonlinearity. Another way to investigate the intensity dependence of the profile is to calculate the effective order k of nonlinearity. It answers the question: For a given frequency, if the ionization probability were given by a power law, what would be the index? It may be defined by

$$k = \frac{I}{\mathcal{N}_1}\frac{d\mathcal{N}_1}{dI} \qquad (4.17)$$

We shall see in Section V, that a realistic calculation of the effective order of nonlinearity should take into account the spatial distribution of the field intensity. However, some important features appear in an analytical formula deduced from Eq. (4.14).

$$k = n + \frac{\Delta_r - \Delta_g}{\delta + \Delta_r - \Delta_g - KL/J} - \frac{2(n-p)L^4}{(\delta + \Delta_r - \Delta_g)^2 + L^4} - \frac{2(\Delta_r - \Delta_g)(\delta + \Delta_r - \Delta_g)}{(\delta + \Delta_r - \Delta_g)^2 + L^4}$$

(4.18)

Far off resonance $k = n$, as expected for a nonresonant process. The second term is only important near the minimum; the divergence it predicts would disappear if several continua were introduced (Edwards and Armstrong, 1980). The third term is a Lorentzian, which describes the depletion of the resonant part of the profile when the intensity is increased. If we remember that this model is most realistic when p is larger than $n - p$, we see that this term is generally small. The last term is in fact that most striking one. It describes a dispersion curve centered near the resonance whose extrema are $\pm(\Delta_g - \Delta_r)/L^2$. In most cases, this ratio is large, possibly several times n. The occurrence of effective order of nonlinearity much larger than n is actually one of the characteristic features of resonant multiphoton ionization. This fact is easily understood from a qualitative description. If the resonance appears for $\delta_0 = \Delta_g - \Delta_r$ at intensity I_0 for a larger detuning $|\delta| > |\delta_0|$ when the intensity is increased, not only does the magnitude of the resonance increase, but the resonance is Stark shifted closer to the field frequency. The greater $(\Delta_g - \Delta_r)/I$, the more important is the phenomenon.

Pulse Shape. Until now we have only dealt with rectangular pulse shape. When an ionization probability may be defined, the number of ions created by a smoothly time-dependent light pulse is calculated as

$$\mathcal{N}_1 = \int_{-\infty}^{+\infty} 2 \frac{[(\delta + \Delta_r - \Delta_g)J - KL]^2}{(\delta + \Delta_r - \Delta_g)^2 + L^4} \, dt$$

(4.19)

In Eq. (4.19) Δ_r, Δ_f, K, L, J now contain a time-dependent intensity $I(t)$. However, even if no experimental light pulse has ever been rectangular, the assumption of a constant field lasting the interaction time accurately describes the most important features of the resonance. Actually, the most critical assumption is the sudden turning on of the field. It is known to be fully justified for on resonance, completely wrong for far off resonance, and questionable for intermediate detuning. Since the location of the resonance varies with the field intensity, it is interesting to compare the effects of a rectangular pulse with those of an adiabatic switching on of the field. For the case of an adiabatic turn on, the state of the system just after the field has been turned on is the unstable eigenstate that evolves from $|g\rangle|N_0\rangle$. That is precisely the state with ionization probability $2\Gamma_-$. Coming back to the discussion of time scale, one sees that once the interaction time is much larger than Γ_+^{-1}, the number of ions created will be the same whenever the switching on of the field is sudden or adiabatic.

2. Resonant State Strongly Coupled to the Initial State

$$K \gg \tfrac{1}{2}L^2 \qquad (4.20)$$

Again we assume that J^2 is much smaller than L^2. In the absence of ionization, the system would exhibit Rabi oscillations described by

$$a_g = \left(\cos \Omega t + i\,\frac{\delta + \Delta_r - \Delta_g}{2\Omega}\,\sin \Omega t\right)\exp\left(-i\,\frac{\delta + \Delta_r + \Delta_g}{2}\,t\right)$$

$$a_r = -i\,\frac{K}{\Omega}\,\sin \Omega t \,\exp\left(-i\,\frac{\delta + \Delta_r + \Delta_g}{2}\,t\right) \qquad (4.21)$$

where Ω is the Rabi frequency defined by

$$\Omega^2 = K^2 + \tfrac{1}{4}(\delta + \Delta_r - \Delta_g)^2 \qquad (4.22)$$

Introducing θ defined by

$$\sin 2\theta = K/\Omega \qquad \text{and} \qquad \cos 2\theta = (\delta + \Delta_r - \Delta_g)/2\Omega \qquad (4.23)$$

the eigenenergies $\lambda_\pm \hbar$ are approximately given by

$$\lambda_+ = \frac{\delta + \Delta_r + \Delta_g}{2} + \Omega - i(L\cos\theta + J\sin\theta)^2$$

$$\lambda_- = \frac{\delta + \Delta_r + \Delta_g}{2} - \Omega - i(L\sin\theta - J\cos\theta)^2 \qquad (4.24)$$

Near the resonance, that is when $|\delta + \Delta_r - \Delta_g|$ is not much larger than K, the ionization probabilities of both eigenstates have the same order of magnitude, about L^2. Hence for interaction times larger than L^{-2}, ionization is saturated, i.e., $\mathcal{N}_i(t) = 1$. For interaction times much smaller than L^{-2}, the ionization probability is a time-dependent function given by

$$\frac{d\mathcal{N}_i(t)}{dt} = 2\left(J^2 \cos^2 \Omega t + \frac{[(\delta + \Delta_r - \Delta_g)J - 2KL]^2}{(\delta + \Delta_r - \Delta_g)^2 + 4K^2}\,\sin^2 \Omega t\right) \qquad (4.25)$$

When the interaction time is much longer than the Rabi period $2\pi\Omega^{-1}$, we obtain a realistic description by averaging the oscillating part of $d\mathcal{N}_i(t)/dt$.

Hence we define a mean ionization probability by

$$\overline{\frac{d\mathcal{N}_1}{dt}} = J^2 + \frac{[(\delta + \Delta_r - \Delta_g)J - 2KL]^2}{(\delta + \Delta_r - \Delta_g)^2 + 4K^2} \tag{4.26}$$

Equation (4.26) describes a Fano profile superimposed on a constant background. As in Section 1, the asymmetry of the profile is related to the interference between resonant and nonresonant quantum paths, contributions to ionization. The contrast of the resonance is now given by $q^2 = L^2/J^2$ and varies as I^{-p}, such that for strong intensities the resonance vanishes. For moderate intensities the profile is almost Lorentzian, centered at

$$\delta = \Delta_g - \Delta_r \tag{4.27}$$

but now the width of the profile is determined by the Rabi frequency (i.e., width $4K$). When the resonant contribution dominates, the effective order of nonlinearity [Eq. (4.17)] is given by

$$k = n - \frac{4pK^2}{(\delta + \Delta_r - \Delta_g)^2 + 4K^2} - \frac{2(\Delta_r - \Delta_g)(\delta + \Delta_r - \Delta_g)}{(\delta + \Delta_r - \Delta_g)^2 + 4K^2} \tag{4.28}$$

The structure of Eq. (4.28) is very similar to the structure of Eq. (4.18): the depletion of the resonance (Lorentzian term) is related to the width of the resonance [$2p$ in Eq. (4.28); $2(n - p)$ in Eq. (4.18)], while the dispersive part is related to the ratio of the shift to the width [$2(\Delta_r - \Delta_j)/L^2$ in Eq. (4.18); $2(\Delta_r - \Delta_g)/(2K)$ in Eq. (4.28)]. The same qualitative explanation holds here.

We shall now discuss how the field is turned on and its effect. When the field is turned on suddenly off resonance, the leading term of the ionization probability [Eq. (4.26)] is

$$\frac{4K^2L^2}{(\delta + \Delta_r - \Delta_g)^2 + 4K^2} \tag{4.29}$$

When the field is turned on adiabatically, the system is in the unstable state with eigenenergy $h\lambda_-$. [Let us define Ω to have the same sign as $(\delta + \Delta_r + \Delta_g)$.] From an expansion of Eq. (4.24), we obtain the ionization probability

$$\frac{2K^2L^2}{(\delta + \Delta_r - \Delta_g)^2 + 4K^2} \tag{4.30}$$

when the field is turned on adiabatically. We thus can expect Eq. (4.26) to give a realistic description of the core of the profile, but the wings are certainly overestimated. Let us remark, finally, that the minimum predicted by Eq. (4.26), is no longer zero and is located at a frequency different from the one predicted by Eq. (4.14). This difference in location of minimum is again related to the use of sudden approximation (Theodosiou et al., 1979).

3. Short Pulse

When $K \ll \frac{1}{2}L^2$, "short" means T is much smaller than L^{-2}; when $K \gg \frac{1}{2}L^2$, it means T is smaller than the Rabi period. We shall now consider the case where the interaction time is much smaller than both K^{-1} and L^{-2}. Equation (4.3) may then be solved by time-dependent perturbation treatment. The number of ions created after a rectangular pulse of duration T is approximately

$$\mathcal{N}_1(T) = 4K^2L^2 \frac{\delta T - \sin \delta T}{\delta^3} - 4KJL \frac{\delta T - \sin \delta T}{\delta^2} + 2J^2T \quad (4.31)$$

The first term is the contribution of resonant quantum paths, the third term corresponds to nonresonant paths, and the second term corresponds to the interference between resonant and nonresonant contributions. The resonant contribution on exact resonance is $4K^2L^2T^3/6$. The width of the resonant profile described by Eq. (4.28) is $7.4/T$. As expected, it is not possible to describe this result with an ionization probability. The number of ions created by a short pulse is characterized by the I^nT^3 dependence of the magnitude of the resonance, which has a width proportional to the reciprocal of the pulse length. The contrast of the resonance is proportional to T^2 and thus decreases when the pulse length is decreased. However, the latter process is difficult to observe experimentally. In most experimental conditions, the contrast of the resonance is so large that it is difficult to detect the resonance and the background for comparable values of the intensity and the pulse duration. Experimental results concerning this situation have been obtained by studying the 4-photon ionization of the cesium ground state, when the 3-photon transition 6s–6f is nearly resonant (Lompré et al., 1978). The variation of the width and the magnitude of the resonance are in good agreement with the results derived here (Crance, 1979; Gontier and Trahin, 1978; Dixit et al., 1980; Andrews, 1977).

A rectangular pulse shape is described by only two parameters: the intensity and the duration. For a smooth time-dependent pulse, especially when nonlinear processes are involved, there are many ways to define the pulse length. For nonresonant processes, it is clear that the relevant duration is τ_n as given by Eq. (2.5). For resonant processes, it is not so simple. This problem reappears in another way when we study short pulses. The condition for a short rectangular pulse is

$$T \ll K^{-1}, L^{-2} \quad (4.32)$$

How is this condition to be transformed when the light pulse is not rectangular? This question and the more general one, the effect of the pulse shape, have been studied by different methods: numerical integration of Eq. (4.3) or approximate analytical treatment for "short" and "long" pulses

(Crance and Feneuille, 1976, 1977; Fedorov, 1977; Fedorov and Kazakov, 1977; Ackerhalt, 1978; Elgin, 1979; Theodosiou *et al.*, 1979; Armstrong and Baker, 1980; Armstrong and O'Neil, 1980; Geltman, 1980; Robinson, 1980). Where short pulses are concerned and the resonance profile is studied, a better definition of condition (4.32) is $\int K(I(t))\,dt \ll 1$ when $K \gg \frac{1}{2}L^2$ and $\int_{-\infty}^{+\infty} L^2(I(t))\,dt \ll 1$ when $K \ll \frac{1}{2}L^2$. In other words, what matters in defining a "short pulse" is the area of the pulse if Rabi oscillations are expected, or the ionization probability of the resonant state if it is weakly coupled to the ground state.

4. Dynamics of Monochromatic Rectangular Excitation

The previous discussion may be summarized in the following way. The dynamics of the process depends essentially on the relative values of

—the probability amplitude K for the p-photon transition g–r,
—the ionization probability $2L^2$ for the resonant state, and
—the interaction time T.

When the ratio $(2K)/L^2$ is either very small or very large, an ionization probability Γ may be defined if the interaction time is long enough

$$(T \gg K^{-1} \quad \text{if} \quad K \gg \tfrac{1}{2}L^2; T \gg L^{-2} \quad \text{if} \quad K \ll \tfrac{1}{2}L^2) \tag{4.33}$$

in which case, the number of ions created is approximately given by $1 - \exp(-\Gamma T)$. For short interaction time, the magnitude of the resonance varies as $I^n T^3$ and the width as T^{-1}. When $(2K)/L^2$ is neither very large nor very small, the characteristic features of short interaction time are expected when T is much smaller than K^{-1} and L^{-2}. Saturation occurs when T reaches the larger of K^{-1} and L^{-2}. However, for intermediate values of T, no time-independent ionization probability may be defined and Eq. (4.5) cannot be simplified.

Another way to discuss Eq. (4.5) is to consider the time dependence of each term in the ionization probability $d\mathcal{N}_1/dt$. Four exponentials appear, two of them oscillating. As we have seen, their relative importance varies with the value of the parameters K, L^2, T. From numerical studies (Ackerhalt and Shore, 1977; Ritchie, 1979; Georges and Lambropoulos, 1977) it is possible to define the range of validity for the approximations studied in Sections IV.A.1 and IV.A.2. The situations where the oscillatory part of $d\mathcal{N}_1(t)/dt$ may be omitted is referred to as a two-rate approximation. When, in addition, one of the remaining exponential terms may be neglected, the situation is referred to as a one-rate approximation. These approximations are valid when the interaction time is long enough. Shorter interaction times may be considered to be a transient regime appearing before the one- or two-rate approximation is valid (Knight, 1979). What happens in this range is, qualitatively, a competition between Rabi oscillation and ionization.

An exhaustive discussion of the possible time dependence of $\mathscr{N}_1(t)$ would be excessively long for this work. However, some amazing features have been predicted for specific situations. A systematic comparison with experiments would certainly be worthwhile in such cases, testing both the dynamics and the quality of the atomic parameters (Geltman, 1980; Haan and Geltman, 1982).

5. Several Resonances

When several quasi-resonant states have to be considered, it is generally not possible to derive an exact analytic formula such as Eq. (4.3). However, it is always possible to derive numerical values for the eigenenergies and ionization probabilities of the unstable states caused by the initial and the resonant states. From these values an equation comparable to Eq. (4.3) may be written (Gontier and Trahin, 1979b).

It is necessary to introduce several resonant states when no transition is exactly resonant, but when several are out of resonance by small but comparable detunings. In such a case, when the field is strong enough, resonant features appear that cannot be attributed to only one of the resonant states, but are owing to the competition of all of them. Then only a calculation involving several resonances will allow one to interpret the ionization probability. For such a situation an adiabatic switching on of the field is usually quite realistic and the ionization probability of the initial state is twice the imaginary part of the eigenenergy obtained for the unstable state arising from $|g\rangle|N_0\rangle$. The eigenenergy of the eigenstate corresponding to $|g\rangle|N_0\rangle$ can be obtained by an iterative procedure (Aymar and Crance, 1979; Crance and Aymar, 1980; Lompré et al., 1980) based on the formula

$$\lambda = \lambda + \frac{\det(H_{\text{eff}} - \lambda\mathbb{1})}{A} \tag{4.34}$$

starting with $\lambda = 0$; where $\mathbb{1}$ is the unit matrix, A is the g–g minor in $(H_{\text{eff}} - \lambda\mathbb{1})$ matrix. The ionization probability is $2\,\text{Im}(\lambda)/\hbar$. The procedure used by Gontier and Trahin to obtain the eigenenergies of the unstable states related to the ground state and the closest resonant state leads them to a continued fraction expansion equivalent to the iterative procedure defined above.

For moderate intensities Eq. (4.34) can be expanded with respect to nondiagonal matrix elements connecting $|g\rangle|N_0\rangle$ to resonant states. This treatment leads to an expression that has exactly the same structure as the ionization probability we would obtain from a conventional perturbative treatment ($\sigma^{(n)}I^n$). However, the energy denominators are different. The relation $E_g + p\hbar\omega - E_r$ appearing in $\sigma^{(n)}$ is replaced by $E_g + \hbar\Delta_g - \hbar iJ^2 - E_r -$

$\hbar\Delta_r + \hbar iL^2$ [see Eq. (4.2)] for each resonant state. This formula has been derived by various methods (McClean and Swain, 1978) and is appropriate only when the coupling between resonant states is weak enough. When two resonant states are strongly coupled (Crance, 1978) or when autoionized states are involved, a lowest-order expansion of Eq. (4.34) does not describe correctly the position and widths of the resonances.

B. Evolution of the Density Matrix

1. Spontaneous Emission

We have defined $\mathbb{P}|\psi(t)\rangle$ as the projection of the wave function on the quasi-resonant states. A density matrix may be defined in the same way. The evolution of this density matrix is given by the Liouville equation:

$$i\hbar \, d\rho/dt = [H_{\text{eff}}, \rho] + \Lambda\rho \qquad (4.35)$$

where H_{eff} is the effective Hamiltonian used to derive Eq. (4.3). The additional term Λ is introduced phenomenologically to include the effects of spontaneous emission. Considering again the case of a single resonance, an explicit expression of Eq. (4.35) is

$$\dot{n}_g = -i(K\alpha_{rg} - K^*\alpha_{gr}) + \gamma_g n_r \qquad (4.36a)$$

$$\dot{n}_r = i(K\alpha_{rg} - K^*\alpha_{gr}) - (2L^2 + \gamma_r)n_r \qquad (4.36b)$$

$$\dot{\alpha}_{gr} = [i(\delta + \Delta_r - \Delta_g) - L^2 - \tfrac{1}{2}\gamma_r]\alpha_{gr} + iK(n_g - n_r) \qquad (4.36c)$$

$$\dot{\alpha}_{rg} = [-i(\delta + \Delta_r - \Delta_g) - L^2 - \tfrac{1}{2}\gamma_r]\gamma_{rg} - iK^*(n_g - n_r) \qquad (4.36d)$$

In a quantum mechanical description ρ is defined by

$$
\begin{aligned}
n_g &= \langle g|\langle N_0|\,\rho|g\rangle|N_0\rangle \\
n_r &= \langle r|\langle N_0 - p|\rho|r\rangle|N_0 - p\rangle \\
\alpha_{gr} &= \langle g|\langle N_0|\rho|r\rangle|N_0 - p\rangle \\
\alpha_{rg} &= \langle r|\langle N_0 - p|\rho|g\rangle|N_0\rangle
\end{aligned}
\qquad (4.37)
$$

In a semiclassical description ρ is defined by

$$
\begin{aligned}
n_g &= \langle g|\rho|g\rangle \\
n_r &= \langle r|\rho|r\rangle \\
\alpha_{gr} &= \langle g|\rho|r\rangle e^{-i\omega t} \\
\alpha_{rg} &= \langle r|\rho|g\rangle e^{i\omega t}
\end{aligned}
\qquad (4.38)
$$

where the field is described by $\frac{1}{2}[\mathscr{E}(t)e^{i\omega t} + \mathscr{E}^*(t)e^{-i\omega t}]$. The quantities K, L^2, Δ_r, Δ_g are the functions of $\mathscr{E}$; K is proportional to $[\mathscr{E}(t)]^p$; Δ_r, Δ_g are proportional to $|\mathscr{E}(t)|^2$; L^2 is proportional to $|\mathscr{E}(t)|^{2(n-p)}$, γ_r is the spontaneous decay rate of state r. Spontaneous emission from r may occur either directly to g or to other states. We are interested in the latter process only when it affects the population of $|g\rangle$. Hence, the term $\gamma_g n_r$ ($\gamma_g < \gamma_r$) stands for any of the spontaneous emission processes that globally lead to populate $|g\rangle$. Let us remark that the particular case where $\gamma_g = 0$ may be studied using the evolution of $\mathbb{P}|\psi(t)\rangle$ by simply replacing L^2 by $L^2 + \frac{1}{2}\gamma_r$.

This oversimplified way of introducing spontaneous emission is justified in the study of multiphoton ionization because of the large magnitude of the light field, but would not be valid for small values of N_0.

The effects of spontaneous emission may be seen by considering how it changes the results of Section IV.A (Theodosiou et al., 1979; McClean and Swain, 1979a,b). When K is much smaller than $\frac{1}{2}L^2$, spontaneous emission hardly affects the ground-state population; the width of the resonance is simply $2L^2 + \gamma_r$, rather than $2L^2$. When K is equal to or much larger than $\frac{1}{2}L^2$, the effect of spontaneous emission is to smooth out the Rabi oscillations. When K is much larger than $\frac{1}{2}L^2$ and the interaction time is much larger than γ_r^{-1}, the competition of Rabi oscillations and spontaneous emission results in a population equilibrium of the $g - r$ two-level system. The ionization probability is then $2L^2\bar{n}_r$, after the transients have died out, where $\bar{n}_r$ is the steady-state population of r. If $\gamma_g = \gamma_r$, the width of the resonance in this case is $(8K^2 + \gamma_r^2)^{1/2}$, instead of $4K$, as obtained in the absence of spontaneous emission.

2. Statistics of the Field

In the previous sections, we have considered a monochromatic excitation, which we assume to be described either quantum mechanically as a number state $|N_0\rangle$ or classically by its amplitude $E(t) = \frac{1}{2}[\mathscr{E}(t)e^{i[\omega t + \phi(t)]} + \text{c.c.}]$, where $\phi(t)$ is constant; $\mathscr{E}(t)$ is real and either constant or a slowly varying function of t. Neither of these descriptions is quite realistic although they provide a good approximation in some cases. In a quantum description, we should represent the field by a density matrix

$$\sum_{\omega, N, N'} \Pi_\omega(N, N')|N\rangle\langle N'|$$

containing a distribution function $\Pi_\omega(N, N')$ for each mode of the field. In a classical description $\mathscr{E}$ and ϕ are often not smooth functions of t; they fluctuate. The assumption of a smooth variation is meaningful only when their fluctuations are slow compared to the dynamics of the process we are considering. Either $\mathscr{E}$ or ϕ or both may fluctuate depending on the characteristics of the source.

In multiphoton ionization experiments, only laser beams are used. The statistics of these laser beams have been studied only for some special cases. For a cw monomode laser well above the threshold with a well stabilized amplitude, it is usually reasonable to assume that only the phase fluctuates. Some short laser pulses, referred to as Fourier transform pulses (Lompré *et al.*, 1978) have a time dependence that is almost the Fourier transform of their spectrum. An approximation commonly used to describe a multimode laser with a large number of uncorrelated modes is an ideal chaotic field (Haken, 1970; Debethune, 1972).

In a classical formalism, the statistics of the field are introduced by assuming that the time variation of $\mathscr{E}(t)e^{i\phi(t)}$ is stochastic. Equations (4.36) are now to be seen as stochastic differential equations that we have to solve for n_g and n_r. The number of ions created will be then given by $1 - \langle n_g \rangle - \langle n_r \rangle$, where $\langle\ \rangle$ stands for the stochastic average. From Eqs. (4.36) we deduce an explicit expression for α_{gr}:

$$\alpha_{gr} = i \int^t dt'\, K(n_g - n_r)$$

$$\times \exp\left\{ \int_{t'}^t \left[i(\delta + \Delta_r - \Delta_g) - L^2 - \tfrac{1}{2}\gamma_r \right] dt'' \right\} \tag{4.39}$$

Taking the stochastic average of Eqs. (4.36a) and (4.36b) gives

$$d/dt\langle n_g \rangle = \gamma_g \langle n_r \rangle - i[\langle K\alpha_{rg} \rangle - \langle K^*\alpha_{gr} \rangle]$$
$$d/dt\langle n_r \rangle = -\gamma_r \langle n_r \rangle + i[\langle K\alpha_{rg} \rangle - \langle K\alpha_{gr} \rangle] \tag{4.40}$$

To solve these equations we need to know $\langle K^*\alpha_{gr} \rangle$ (and $\langle K\alpha_{rg} \rangle$), which we can obtain from Eq. (4.39) by multiplying by K^* and taking the stochastic average, i.e.,

$$\langle K^*\alpha_{gr} \rangle = i \int^t dt'\, \langle K^*(t) \Big| \Big\langle (t')(n_g(t') - n_r(t'))$$

$$\times \exp\left[\int_{t'}^t \left[i(\delta + \Delta_r - \Delta_g) - L^2 - i\tfrac{1}{2}\gamma_r \right] dt'' \right] \Big\rangle \tag{4.41}$$

Generally, the stochastic average appearing in the right-hand side of Eq. (4.41) is not simply related to $\langle n_g \rangle - \langle n_r \rangle$. The point in studying the effect of field statistics is to find a way to reduce Eq. (4.41) to a tractable function of $\langle n_g \rangle$, $\langle n_r \rangle$, and the field characteristics. The solution depends on how the field statistics is modeled. So far solutions have been found only in a few cases. (Lambropoulos, 1974; Kovarskii and Perelman, 1975; Ackerhalt and Eberly, 1976; Armstrong *et al.*, 1976; Mostowski, 1976; Wong and Eberly, 1977; Agostini *et al.*, 1978; Dixit and Lambropoulos, 1978, 1980; Georges and Lambropoulos, 1978; de Meijere and Eberly, 1978; Sanchez, 1978; Gontier and Trahin, 1979a; Swain, 1979; Zoller and Lambropoulos, 1979; Zoller, 1979; Yeh and Eberly, 1981). Actually, the situations investigated may be summarized in the following way. When phase fluctuations occur, the light is not monochromatic and may be characterized by the intensity spectrum. The spectrum is always assumed to be Lorentzian near the central frequency. However, some authors propose a more realistic model of the wings which actually fall off more rapidly than a Lorentzian dependence predicts. When amplitude fluctuations are considered, the field is assumed to be chaotic. Hence, what is discussed is the comparison between constant and chaotic amplitude for various bandwidths of the light spectrum. The complexity of an accurate solution depends on the magnitude of the relevant parameters describing the atom and the field. We shall discuss the simplest cases and only give an idea of the more complex treatments.

Phase Fluctuations. When only phase fluctuations exist, the exponential appearing in Eq. (4.41) factors out since it only depends on the intensity of the field. The remaining term is

$$\langle K^*(t)K(t')[n_g(t') - n_r(t')]\rangle$$

which in fact reduces to

$$|K|^2\langle[n_g(t') - n_r(t')] \exp\{ip[\phi(t') - \phi(t)]\}\rangle \qquad (4.42)$$

Various descriptions of the phase fluctuations have been used, the simplest one assuming a Wiener–Levy process. This model is generally referred to as the phase diffusion model. This model has been widely used for a variety of problems. It has been shown that in most situations, products such as Eq. (4.42) may be decorrelated (Wodkiewicz, 1979), that is, written as

$$\langle(n_g(t') - n_r(t'))\rangle\langle\exp\{ip[\phi(t') - \phi(t)]\}\rangle \qquad (4.43)$$

In the phase diffusion model, the two time correlation functions are given by

$$\langle\exp\{ip[\phi(t') - \phi(t)]\}\rangle = \exp[-p^2\beta|t - t'|] \qquad (4.44)$$

(Agarwal, 1978). Under this assumption, Eq. (4.41) reduces to

$$\langle K^*\alpha_{gr}\rangle = i\int_0^t dt'(\langle n_g\rangle - \langle n_r\rangle)|K|^2$$
$$\times \exp[(i(\delta + \Delta_r - \Delta_g) - L^2 - \tfrac{1}{2}\gamma - p^2\beta)(t - t')] \qquad (4.45)$$

Differentiating Eq. (4.45), we obtain a system of equations similar to the initial one [Eqs. (4.36)], but now governing stochastic averages:

$$\frac{d}{dt} \langle n_g \rangle = \gamma_g \langle n_r \rangle - i \left[\frac{\langle K \alpha_{rg} \rangle}{|K|} - \frac{\langle K^* \alpha_{gr} \rangle}{|K|} \right] |K| \tag{4.46a}$$

$$\frac{d}{dt} \langle n_r \rangle = -\gamma_r \langle n_r \rangle + i \left[\frac{\langle \alpha_{rg} \rangle}{|K|} - \frac{\langle K^* \alpha_{gr} \rangle}{|K|} \right] |K| \tag{4.46b}$$

$$\frac{d}{dt} \frac{\langle K^* \alpha_{gr} \rangle}{|K|} = [i(\delta + \Delta_r - \Delta_g) - L^2 - \tfrac{1}{2}\gamma_r - p^2 \beta] \frac{\langle K^* \alpha_{gr} \rangle}{|K|}$$

$$+ i|K|(\langle n_g \rangle - \langle n_r \rangle) \tag{4.46c}$$

This differs from Eqs. (4.36a)–(4.36d) only by the addition of a damping term in Eq. (4.46c). Qualitatively, the effect of this damping term is to smooth the Rabi oscillations and to eventually broaden the resonance profile.

Phase-Amplitude Fluctuations. The intensity-dependent terms such as Δ_r, Δ_g, L^2 are now stochastic functions of time. In Eq. (4.41) we have to estimate the average:

$$\left\langle K^*(t) K(t') (n_g(t') - n_r(t')) \exp\left[\int_{t'}^t [i(\delta + \Delta_r - \Delta_g) - L^2 - \tfrac{1}{2}\gamma_r] dt'' \right] \right\rangle$$

The fluctuations are characterized by a coherence time T_c. When the mean values of the frequencies appearing in the Liouville equations are smaller than T_c^{-1}, it is still a good approximation to decorrelate the population term $\langle n_g \rangle - \langle n_r \rangle$, expand the exponential, and express it in terms of the correlation functions of the field. Although this method appears to give realistic predictions for rather larger intensities, it is really a weak field approximation. This method has been applied to study the 2-photon resonant, 3-photon ionization of sodium (Agostini *et al.*, 1978), where the field is assumed to be an ideal chaotic field characterized by the correlation function

$$\langle \mathscr{E}(t_1) \mathscr{E}^*(t_2) \rangle = \mathscr{E}_0^2 \exp[-\beta |t_1 - t_2|] \tag{4.47}$$

Here, $\mathscr{E}_0^2$ is the mean intensity and β^{-1} is the correlation time. Under these conditions, Eq. (4.46c) is replaced by

$$\frac{d}{dt} \frac{\langle K^* \alpha_{gr} \rangle}{|K_0|} = [i(\delta + 3(\Delta_g)_0 - 3(\Delta_r)_0) - \tfrac{1}{2}\gamma_r - 3(L^2)_0 - 4\beta] \frac{\langle \alpha_{gr} K^* \rangle}{|K_0|}$$

$$+ 2i|K_0|(\langle n_g \rangle - \langle n_r \rangle) \tag{4.48}$$

$(L^2)_0$, $(\Delta_r)_0$, $(\Delta_g)_0$, K_0 are the values of L^2, Δ_r, Δ_g, K when $\mathscr{E}$ is replaced by $\mathscr{E}_0$.

Once again, an additional damping to the nondiagonal density matrix element (4β) appears but additional modification must also be noted. A factor of 2 in front of the population difference corresponds to the well-known enhancement of $\sqrt{p!}$ in a nonresonant multiphoton transition amplitude (here $p = 2$). Moreover, the ionization damping $(L^2)_0$ and the shift $(\Delta_g)_0$ and $(\Delta_r)_0$ are now multiplied by a factor of 3. From this model a complete calculation of the number of ions created $\mathcal{N}_1$ has been performed and the result is illustrated by Fig. 2, where $\mathcal{N}_1$ is plotted as a function of intensity for two different bandwidths of the light field. (Relative numbers of ions were measured in the experiment and a single scaling factor was used to make the comparison between experimental data and theoretical predictions.)

The assumption of Eq. (4.47) corresponds to a Lorentzian spectrum for the field. Such a spectrum is reasonably realistic near the central frequency, but the intensity probably falls off faster than a Lorentzian in the far wings. To improve the model of Eq. (4.47), more complex correlation functions have been studied (Zoller and Lambropoulos, 1979; Yeh and Eberly, 1981). We shall not discuss the result in detail since it affects essentially the far wings of the resonance curve.

Most works have emphasized the effect of light statistics on the resonant ionization probability. As we shall stress in the next section, it is necessary to know accurately the time evolution of the light shifts in order to compare

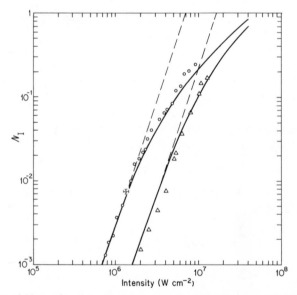

Fig. 2. Ion yield as a function of laser intensity for two different bandwidths. [Reprinted with permission from Agostini et al. (1978). Copyright 1978 by The Institute of Physics.]

theoretical calculations with experimental data. It has been shown, for example, that amplitude fluctuations of the field produce an enhancement of the ac-Stark shift. This was demonstrated in a study of 3-photon ionization of Na (Zoller and Lambropoulos, 1980). Results showing the Stark shift are presented in Fig. 3, where the predictions for a chaotic field are plotted as a function of intensity for two pulse lengths. These results are also compared with the result of the phase diffusion model (no enhancement) and the decorrelation assumption.

The weakness of the treatment outlined above is the assumption that the population terms $\langle n_g - n_r \rangle$ and the field-dependent terms can be decorrelated. Different treatments have been proposed which avoid this assumptions, thus allowing study of a large range of parameters both for the light pulse and the atom.

Zoller (1979) considers a Markovian chaotic field represented by its Fokker–Planck equation. The stochastic density matrix equations are reduced to an infinite set of differential equations for the required stochastic averages. This system, suitably truncated, is solved numerically. Georges and Lambropoulos (1979) treat directly the stochastic density matrix equations. A series expansion is obtained in a diagramatic form and the field-dependent terms are expressed in terms of the correlation functions of the field. Analytic expressions are obtained in some special cases. The most striking results of these studies may be summarized by considering the probability $P(t)$ for an atom to be ionized at time t. For very small values of $P(t)$, such as would result from short interaction times or weak intensities, one finds the chaotic result is enhanced by $N!$ over the coherent result, just as predicted by perturbation theory. As soon as $P(t)$ is not very small, a purely coherent field

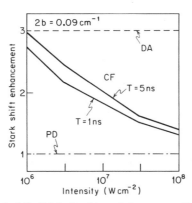

Fig. 3. Predicted Stark shift. DA is the decorrelation assumption, curve PD is the phase diffusion model, curves CF for a chaotic field, when the interaction time is T. [Reprinted with permission from Zoller and Lambropoulos (1980). Copyright 1980 by The Institute of Physics.]

(no phase or amplitude fluctuations) is more efficient than a monochromatic chaotic field (no phase fluctuations). A finite bandwidth is less effective than a monochromatic chaotic field in most cases. However, for long interaction times, there may exist a nonzero optimum bandwidth.

V. SPATIAL DISTRIBUTION OF INTENSITY

In order to calculate the number of ions created by a light pulse, we can usually use the results of the previous section for an intensity $I(\mathbf{r})$ and integrate the contributions over the interaction volume. Such a calculation preserves most of the characteristic features explained in Section IV, and the analytical results of this section can often be directly compared to experimental data as far as relative values of ion yield are concerned. However, in some cases the role of the spatial distribution of intensity is crucial and the complicated procedure described above has to be used. For example, the effective order of nonlinearity is extremely sensitive to the inhomogeneity of the spatial distribution of intensity. The number of ions created is not so sensitive to the distribution of intensity when the ionization probability per light pulse remains small. However, when saturation occurs for the maximum intensity in the distribution, it is necessary to take into account the spatial distribution of intensity in order to obtain a reliable description of the number of ions created.

Effective Order of Nonlinearity. The total number of ions created may be written as

$$\mathcal{N}_T(I_M) = \mathcal{N}_0 \int d\mathbf{r}\, \mathcal{N}_i[I(\mathbf{r})] \tag{5.1}$$

where $\mathcal{N}_i(I)$ refers to the results of Section IV. The integration in (5.1) is taken over the interaction volume. $\mathcal{N}_0$ is the number of atoms per unit volume and I_M is the maximum value of the intensity. Let us consider, in order to have a qualitative discussion, the case where an ionization probability may be defined. Schematically $\mathcal{N}_i(I)$ describes a resonance centered at the detuning $\Delta_g(I) - \Delta_r(I)$, with a magnitude proportional to I^{2p-n} (or I^{n-p}) and a width proportional to I^{n-p} (or $I^{p/2}$). If the spatial distribution of intensity falls off rapidly, the largest contribution to the ionization comes from the volume where I has almost its maximum value I_M; then $\mathcal{N}_T(I_M)$ will have a maximum for detunings close to $\delta_M = \Delta_g(I_M) - \Delta_r(I_M)$. For detunings δ smaller than δ_M, the process is still resonant for the atoms receiving an intensity I such that $\delta = \Delta_g(I) - \Delta_r(I)$. However, for detunings larger than δ_M, the process is not resonant for any atom, and the main contribution to the ionization probability is from the atoms receiving the maximum intensity. The result for $\mathcal{N}_T(I_M)$ is an asymmetric shape, similar to $\mathcal{N}_i(I_M)$ for large detunings. Actually, the shape of $\mathcal{N}_T(I_M)$ differs strongly from the shape of $\mathcal{N}_1(I_M)$ when the light shift of the p-photon resonance is larger than the width of the resonance; that is, when

$$\left|\frac{\Delta_g(I_M) - \Delta_r(I_M)}{\Gamma(I_M)}\right| \gg 1 \tag{5.2}$$

$\Gamma(I_M)$ is the larger of $L^2(I_M)$ and $K(I_M)$. Figure 4 shows the profile predicted for $\mathcal{N}_T$ as a function of the field frequency for various values of the shift over width ratio. A Gaussian distribution of intensity has been assumed.

The resulting effective order of nonlinearity can be recalculated as

$$k_I = \frac{I_M}{\mathcal{N}_T}\frac{d\mathcal{N}_T}{dI_M} \tag{5.3}$$

The large maximum predicted by Eq. (4.18) and (4.28) occurs for detunings larger than δ_M. As the shape of $\mathcal{N}_T(I_M)$ is similar to the shape of $\mathcal{N}_1(I_M)$ for detunings larger than δ_M, the maximum of k_1 will hardly differ from the maximum predicted by Eq. (4.18) and (4.28). However, the large minimum predicted in Eqs. (4.18) and (4.28) vanishes and is replaced by a shallow wide minimum which appears for detunings intermediate between 0 and δ_M. The negative values of k predicted by Eqs. (4.18) and (4.28) disappear when a Gaussian distribution of intensity is assumed (see Fig. 4b). Negative values might, however, appear for a distribution of intensity falling more rapidly

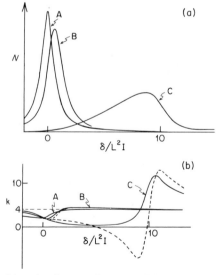

Fig. 4. (a) Predicted ion yield [Eq. (5.1)] and (b) effective order of nonlinearity [Eq. (5.3)] for various values of the shift over width ratio. [Reprinted with permission from Crance (1980). Copyright 1980 by The Institute of Physics.]

than in a Gaussian beam. In fact, a decrease of $\mathcal{N}_T$ when I_M is increased has been observed experimentally (Baravian *et al.*, 1976).

Of theoretical interest is k_I, but it is generally not measured in a large range of frequency. In most experiments, it is possible to vary the maximum light intensity in a large range, but the number of ions created may be detected only in a narrow range. The lower limit of detection is defined by the sensitivity of the apparatus, the higher limit may be determined by the saturation, but also by the appearance of new processes such as space charge. As a result, the experimental data generally describe the ionization when the number of ions created is constant or varies in a limited range (a few orders of magnitude). The experimental effective order of nonlinearity is thus better described by

$$k_N = \frac{I_M}{\mathcal{N}_T}\left(\frac{d\mathcal{N}_T}{dI_M}\right)_{\mathcal{N}_T} \tag{5.4}$$

the derivative being taken, not for a constant value of the intensity as in Eq. (4.18) and (4.28), but for a constant value of $\mathcal{N}_T$. Like k_I, k_N, has a dispersion shape, the maxima are comparable (in fact equal when the resonant state is ionized by absorption of one photon). However, the maximum and the minimum of k_N are obtained for weak intensities and thus are located near the zero field resonance $\delta = 0$. The difference of shape between k_N and k_I is also illustrated in Fig. 5, where experimental results are compared with theoretical calculations using Eqs. (5.3) and (5.4).

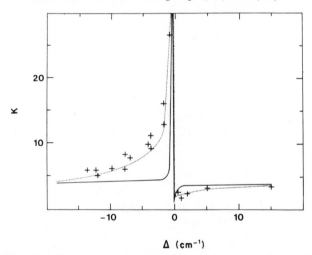

Fig. 5. Theoretical effective order of nonlinearity compared to experimental data. Full line from Eq. (5.4). [Reprinted with permission from Gontier and Trahin (1980). Copyright 1980 by The Institute of Physics.]

Saturation Near a Resonance. For a given pulse shape, we may define the saturation intensity I_S as the maximum intensity (with respect to time) of a light pulse for which the ionization probability per pulse is of the order of 1. For maximum intensities (with respect to space) much larger than I_S, most atoms are ionized and the number of ions created in the interaction volume is about $\mathcal{N}_0 \mathscr{V}(I_S)$, $\mathscr{V}(I_S)$ being the volume where the intensity is larger than I_S. Describing $\mathcal{N}_T(I_M)$ for values of I_M beyond the saturation intensity requires a precise knowledge of the intensity distribution of the field, especially for the lowest relative intensities. For nonresonant multiphoton ionization, the number of ions created varies as I_M^n for intensities weaker than I_S; when the intensity becomes closer to I_S, the effective order of nonlinearity decreases until the saturation regime described by $\mathcal{N}_0 \mathscr{V}(I_M)$ is reached. Near, but not on, a resonance, $\mathcal{N}_T(I_M)$ still varies as I_M^n in a weak field. When the intensity becomes so large that the resonant processes are important, $\mathcal{N}_T(I_M)$ varies as I_M^k, where k differs from n; a competition between resonance effects and saturation may occur. If k is larger than n, saturation seems to be delayed: the increase of ionization probability compensates the depletion owing to saturation. If k is smaller than n, the saturation intensity may be overestimated because the decrease of ionization probability cannot be distinguished from saturation effects. In such cases, only a precise calculation incorporating a correct description of the spatial distribution of intensity enables one to distinguish the effect of saturation from the effect of resonance.

An example of the first situation $(k > n)$ is provided by the study of 4-photon ionization of Cs at frequency $\omega = 9470 \text{ cm}^{-1}$ (Morellec and Normand, 1979). The 3-photon transition 6s–6f is resonant in weak field for $\omega = 9443 \text{ cm}^{-1}$ and when the intensity is increased, the resonance frequency is shifted towards higher frequencies. As a result, the number of ions created increases slightly faster than I_M^4. However, saturation occurs before the departure from a I_M^4 law is strong enough to be noticeable. An example of the situation where $k < n$ is provided by the study of 3-photon ionization of the $(1s2s)^1 S$ metastable state of He at $\omega = 14398.5 \text{ cm}^{-1}$ (Lompré *et al.*, 1980). The 2-photon transition $(1s2s)^1 S$–$(1s6s)^1 S$ is resonant in weak field for $\omega = 14419 \text{ cm}^{-1}$. When the intensity is increased, the resonance frequency is shifted towards higher frequencies. As a result, the number of ions created increases slower than I_M^3. Again, the range of intensity where the measurement was possible was too narrow for a departure from a power law to be observable. In such situations, the meaning of the atomic quantities deduced from experimental measurements may be ambiguous, since the intensity dependent ionization cross section $(P(I_M)/I_M^n)$ which is measured may differ noticeably from the weak field cross section $\sigma^{(n)}$ (Aymar and Crance, 1979).

VI. CALCULATION OF ATOMIC QUANTITIES

Our purpose in this section is to present methods for calculating atomic quantities that can be used in making comparisons between theory and experiment. We shall first describe methods for calculating the effective Hamiltonian discussed in Section III. When an ionization probability may be defined, one can, of course, calculate it directly without introducing an effective Hamiltonian, but for the sake of generality we shall proceed via an effective Hamiltonian. In studying the dynamics of the process we have not previously discussed the effects of the polarization of the field; we shall do so in this section.

A. Calculation of an Effective Hamiltonian Structure and Its Dependence on Light Polarization

The structure of the effective Hamiltonian derived in Section III is very similar to the structure of generalized cross sections [$\sigma^{(n)}$ Eq. (2.2)] encountered in the description of nonresonant processes. Hence the same general methods can be used to calculate both quantities. In our description of these methods, we shall stress the differences between resonant (effective Hamiltonian) and nonresonant (generalized cross section) situations.

The quantities to calculate can be written in the general form

$W_{i; \, N_0 - p_i; \, j, \, N_0 - p_j}$

$$
= \sum_{\alpha, \beta, \ldots, \gamma} \frac{\langle i | \langle N_0 - p_i | V | \alpha \rangle | N_0 - p_\alpha \rangle \langle \alpha | \langle N_0 - p_\alpha | V | \beta \rangle | N_0 - p_\beta \rangle}{[E_0 - (N_0 - p_\alpha)\hbar\omega - E_\alpha][E_0 - (N_0 - p_\beta)\hbar\omega - E_\beta]} \frac{\cdots \langle \gamma | \langle N_0 - p_\gamma | V | j \rangle | N_0 - p_j \rangle}{\cdots [E_0 - (N_0 - p_\gamma)\hbar\omega - E_\gamma]}
$$

$$(6.1)$$

What the states $|i\rangle |N_0 - p_i\rangle$, $|j\rangle |N_0 - r_j\rangle$ may be and the range of the summations are different in the resonant and nonresonant situation. In the nonresonant case $|i\rangle |N_0 - p_i\rangle$ stands for $|g\rangle |N_0\rangle$ and $|j\rangle |N_0 - p_j\rangle$ stands for any of the resonant continua. The intermediate summations run over all the intermediate states of the atom field. Morever, if we write the interaction operator V as $V = V^+ + V^-$, with

$$V^\pm = (2\hbar\omega/\varepsilon_0 v)^{1/2} \, da^\pm$$

[see Eq. (3.4)], only the V^+ terms contribute. In the resonant situation $|i\rangle |N_0 - p_i\rangle$ and $|j\rangle |N_0 - p_j\rangle$ stand for any of the states spanning $\mathscr{E}_R$ or any resonant continuum. The intermediate summations run over all the states except the ones spanning $\mathscr{E}_R$, and there are terms in which both V^+

and V^- contribute. If we rewrite an expression such as Eq. (6.1) in terms of V^+ and V^-, each term of the sum can be interpreted as describing a quantum path going from $|i\rangle|N_0 - p_i\rangle$ to $|j\rangle|N_0 - p_j\rangle$ corresponding to successive absorption and emission of a photon. The quantum path describing the H_{eff} matrix element between $|g\rangle|N_0\rangle$ and $|j\rangle|N_0 - p_i\rangle$ corresponds to an absorption of p_i photons. For any diagonal matrix element two quantum paths appear; one corresponds to an absorption followed by an emission, the other one corresponds to an emission followed by an absorption. The same two types of quantum paths will appear in the H_{eff} matrix element between two quasi-resonant states belonging to the same multiplicity ($|i\rangle|N_0 - p_i\rangle$ and $|j\rangle|N_0 - p_j\rangle$ when $p_i = p_j$).

The next step is to split each term in an angular part and a radial part. The atomic dipole operator is $d = \mathbf{r} \cdot \mathbf{e}$, where $\mathbf{e}$ is the polarization of the field. For a linearly polarized light $d = rC_0^{(1)}$, and for a circularly polarized light $d = rC_{\pm 1}^{(1)}$, where $C_q^{(1)}$ is a renormalized spherical harmonic

$$C_q^{(l)} = (4\pi/2l + 1)^{1/2} Y_{lq}.$$

The effect produced by the field polarization results from the selection rules on matrix elements of $C^{(1)}$. Let an atomic state be characterized by its total momentum J and its projection M. For any polarization of light, the selection rule is $\Delta J = 0, \pm 1$ between two successive intermediate summations. If the initial state is $J_0 M_0$ all the states involved in intermediate summations have $M = M_0$ if linearly polarized light is used. When the light is circularly polarized with $q = +1$, $\Delta M = +1$ for each step involving absorption of a photon, $\Delta M = -1$ for each step involving emission of a photon. (The opposite relationship holds for $q = -1$.) If we now introduce the more elementary quantum paths $J_0 M_0, J_1 M_1, \ldots$ to describe the effective Hamiltonian matrix elements, the difference between circular and linear polarization is that some quantum paths $J_0 J_1, \ldots$ which are allowed in linear polarization are forbidden in circular polarization because of the selection rules on ΔM. Correspondingly, some states are resonant in linear polarization and not in circular polarization. Moreover, all the resonant continua are reached in linear polarization; only some of them are reached in circular polarization. This difference is important because it allows one to overcome, to some extent, the problem of absolute measurements. The ratio of the ion yield in circular and linear polarization is often measured more accurately than the ion yield itself. This ratio depends on the relative values of the contribution of each quantum path $J_0 J_1, \ldots$. For a given path the radial part is the same, the angular part is different for different polarizations. For nonresonant multiphoton ionization, it is thus possible to assign an upper limit to this ratio that is $(2n - 1)!!/n!$ for n-photon ionization. For resonant multiphoton ionization, the atomic parameters will be different for circular and linear polarization, leading to a time evolution that is different in the two cases. Thus the measured ratio may be larger than the limit predicted by the nth order perturbative treatment. (Dixit, 1981; Parzynski, 1979, 1980b.)

Finally, we note that, owing to the selection rules in circular polarization, the photoelectron may be polarized (Lambropoulos, 1973; Lambropoulos and Teague, 1976; Parzynski, 1980a). In addition, if hyperfine structure exists, polarized nuclei may be produced by resonant multiphoton ionization (Parzynski, 1980c). When the vapor is a mixture of several isotopes, both hyperfine and isotropic structure may exist and resonant multiphoton ionization is more efficient for certain isotopes, depending on the field frequency (Letokhov and Mishin, 1979).

B. Calculation Using the Dalgarno Method

The angular part of each term may be calculated by using the standard techniques of angular momentum algebra. The remaining radial parts consist of monoelectronic expressions such as

$$\sum_{n_1 l_1 \cdots n_k l_k} \frac{\langle n_0 l_0 | r | n_1 l_1 \rangle \langle n_1 l_1 | r | n_2 l_2 \rangle \cdots \langle n_k l_k | \gamma | n_{k+1} l_{k+1} \rangle}{(E_0 + \hbar\omega - E_1)(E_0 + 2\hbar\omega - E_2) \cdots (E_0 + k\hbar\omega - E_k)} \quad (6.2)$$

when a $(k + 1)$th order element is calculated between state $|0\rangle$ and state $|k + 1\rangle$. Here $E_0, E_1, E_2, \ldots, E_{k+1}$ are the energies of the states $|0\rangle, |1\rangle, \ldots, |k + 1\rangle$, and $n_0 l_0, n_1 l_1, \ldots$ are monoelectronic states appearing in the expansion of the states $|0\rangle, |1\rangle, \ldots, |k + 1\rangle$ in a configuration basis. Since all the states $|1\rangle, |2\rangle, \ldots, |k\rangle$ are far from resonance, the energy denominators will not change too much if we replace $E_1, E_2, \ldots$ by the corresponding monoelectronic energies $E_{n_1 l_1}, E_{n_2 l_2}$. The Dalgarno method is a calculation step by step by expressions such as Eq. (6.2). Let us consider the function

$$|f_k\rangle = \sum_{n_k} \frac{|n_k l_k\rangle \langle n_k l_k | r | n_{k+1} l_{k+1} \rangle}{E_0 + k\hbar\omega - E_{n_k l_k}} \quad (6.3)$$

Let h_{lk} be the radial part of the monoelectronic Hamiltonian for angular momentum l_k; then

$$h_{l_k} |n_k l_k\rangle = \mathscr{E}_{n_k l_k} |n_k l_k\rangle \quad (6.4)$$

The function $|f_k\rangle$ then satisfies the inhomogeneous differential equation

$$(E_o + k\hbar\omega - h_{lk})|f_k\rangle = r|n_{k+1} l_{k+1}\rangle \quad (6.5)$$

The next intermediate function

$$|f_{k-1}\rangle = \sum_{n_k n_{k-1}} \frac{|n_{k-1} l_{k-1}\rangle \langle n_{k-1} l_{k-1} | r | n_k l_k \rangle \langle n_k l_k | r | n_{k+1} l_{k+1} \rangle}{(E_0 + (k-1)\hbar\omega - E_{n_k l_k})(E_0 + k\hbar\omega - E_{n_k l_k})} \quad (6.6)$$

satisfies the differential equation

$$(E_0 + (k-1)\hbar\omega - h_{l_{k-1}})|f_{k-1}\rangle = r|f_k\rangle \tag{6.7}$$

This description holds for nonresonant processes (Dalgarno and Lewis, 1955). If a resonant state has to be removed from the summation over k states, then a corresponding orbital $n_k l_k$ is to be removed from the corresponding summation in Eq. (6.2) and Eq. (6.5) becomes

$$(E_0 + k\hbar\omega - h_{lk})|f_k\rangle = r|n_{k+1}l_{k+1}\rangle - |n_k l_k\rangle\langle n_k l_k|r|n_{k+1}l_{k+1}\rangle \tag{6.8}$$

Each of the successive inhomogeneous differential equations must be modified in the same way.

We have not specified the exact form of h_{lk} since it depends on the atomic model chosen to calculate the wave functions. The central potential approximation leads to a particularly simple form of h_{lk} (Crance and Aymar, 1979, 1980; Aymar and Crance, 1979; Lompré et al., 1980; Chang and Poe, 1976, 1977), but more complicated potentials such as the Hartree–Fock have also been used (Pindzola et al., 1981).

In the same spirit as the Dalgarno method, a solution has been proposed to calculate ionization probabilities in a nonperturbative way (Gontier et al., 1981). In a few words, it is a method to solve directly the hierarchy of differential equations appearing in the Dalgarno method in a nonperturbative way. However, it has not been applied to any experimental case yet.

C. Calculation of Ionization Probabilities by Using Green's Function

An alternative to the Dalgarno method for performing the infinite summations which appear in Eq. (6.1) is to introduce the Green's functions associated with the Hamiltonian. This has been done only when the atomic wave functions are known analytically, as in the case of the hydrogen atom (Ritchie, 1978, 1979), or in the framework of the quantum defect theory (Declemy et al., 1981; McGuire, 1981). For a Hamiltonian with a complete basis of discrete states $|\alpha\rangle$ with energy E_α, and continuum states $|\varepsilon\rangle$ with energy E_ε, the Green's function $\mathbb{G}(E, \mathbf{r}, \mathbf{r}')$ may be defined by

$$\mathbb{G}(E, \mathbf{r}, \mathbf{r}') = \sum_\alpha \frac{\langle \mathbf{r}|\alpha\rangle\langle\alpha|\mathbf{r}'\rangle}{E - E_\alpha} + \int d\varepsilon \frac{\langle \mathbf{r}|\varepsilon\rangle\langle\varepsilon|\mathbf{r}'\rangle}{E - E_\varepsilon} \tag{6.9}$$

$\mathbb{G}(E, \mathbf{r}, \mathbf{r}')$ is nothing but the resolvent operator matrix expressed in the basis $|\mathbf{r}\rangle$. Let us consider the example of a second-order term appearing in perturbation treatment such as

$$A = \sum_{\alpha,\varepsilon} \frac{\langle 0|d|\alpha\rangle\langle\alpha|d|1\rangle}{E_0 + \hbar\omega - E_\alpha} \tag{6.10}$$

It may be rewritten more explicitly by introducing the closure relation $\int |\mathbf{r}\rangle\langle\mathbf{r}| \, dr$ twice.

$$A = \iint dr \, dr' \, \langle 0|\mathbf{r}\rangle \, d(\mathbf{r}) \, d(\mathbf{r}')\langle\mathbf{r}'|1\rangle \left(\sum_{\alpha,\varepsilon} \frac{\langle\mathbf{r}|\alpha\rangle\langle\alpha|\mathbf{r}'\rangle}{E_0 + \hbar\omega - E_\alpha} \right) \qquad (6.11)$$

The quantity in brackets is the Green's function for $E = E_0 + \hbar\omega$ and it may be rewritten

$$A = \iint dr \, dr' \, \langle 0|\mathbf{r}\rangle d(\mathbf{r}) d(\mathbf{r}')\langle\mathbf{r}'|1\rangle \mathbb{G}(E_0 + \hbar\omega, \mathbf{r}, \mathbf{r}') \qquad (6.12)$$

In this formalism, as in the Dalgarno method, the angular part factorizes and reduced Green's functions $g_l(E, r, r')$ depending only on radii are defined by

$$\mathbb{G}(E, \mathbf{r}, \mathbf{r}') = \sum_l g_l(E, r, r') Y_{lm}(\theta, \phi) Y_{lm}(\theta', \phi') \qquad (6.13)$$

when the coordinates r, θ, ϕ, define $\mathbf{r}$ and r', θ', ϕ' define $\mathbf{r}'$. In this framework, the quasi-resonant states are removed from infinite summation by a suitable modification of the corresponding Green's function. By using an iterative method similar to the procedure we described in Section IV, Ritchie studies multiphoton ionization of hydrogen in the vicinity of resonances. The intensity dependent ionization probabilities calculated by this method may be used directly when the assumption of an adiabatic turn on of the field is valid. When this approximation breaks down, and particularly near resonance, the same calculation may be done for the quasi-resonant states, in order to allow a complete calculation of the dynamics to be carried out.

D. Direct Calculation of Ionization Probabilities— Diagonalization of the Complete Hamiltonian

Off resonance, the ionization probability of the initial state is twice the imaginary part of the pseudoenergy of the corresponding state of $H_0 + V$. This statement is valid in two situations (see Section IV); when the field is turned on adiabatically whatever the interaction time is; and when the field is turned on suddenly but:

1. The coupling between the initial state and the closest resonant state is weaker than the coupling between the resonant state and the ionization continuum.

2. The interaction time is such that the transients have already died out but the ionization is not saturated yet. If it were possible to write down the complete Hamiltonian matrix for $H_0 + V$ and diagonalize it, then the ionization probability of any state would be obtained. It is not possible because the atomic basis ($|i\rangle$) and the field basis $|N\rangle$ are infinite. However, the pre-

vious analysis in Section III shows that the initial state $|g\rangle|N_0\rangle$ is significantly coupled only to a few field multiplicities $|N\rangle$, hence we can expect that truncating the field state basis to some multiplicities such that $N_0 - m_- < N < N_0 + m_+$ will give a good approximation. (Actually the stability of the result may be checked by increasing $m_\pm$ step by step.) Although the atomic basis is infinite, it is possible, at least for hydrogen, to find a finite basis which well represents the atomic states. After these truncations and replacements, it is possible to write down the Hamiltonian $H_0 + V$ for any field intensity in a finite basis. This method has been used by Chu and Reinhardt (1977) to study the 2-photon ionization of the ground state of hydrogen. Instead of searching for the eigenvalues of the Hamiltonian, they search for the pole of the resolvent. They are able to study intensities up to $7. \times 10^{12}$ W cm^{-2} by introducing at most five multiplicities. This method is certainly the simplest and the most powerful one among the calculations already performed. However its applicability to other atoms than hydrogen has not been proven.

VII. ELECTRON YIELD

In the previous sections, we have studied the probability for an atom to be ionized within a particular interaction time. These results are useful in the interpretation of experiments in which either the total number of ions or the total number of electrons created is measured. However it is possible to obtain additional information by studying either the angular distribution or the energy spectrum of the photoelectrons. The relevant atomic quantities can be obtained from the formalism described above. An electron state, considered as a state belonging to the atomic continuum may be characterized by its energy ε, orbital momentum l, and its projection m along the quantization axis (or total angular momentum $\mathbf{j} = \mathbf{l} + \mathbf{s}$ if we are interested in spin polarization). We shall refer to such a continuum state by $|\varepsilon, \alpha\rangle$, where α is the set of quantum numbers required to label energy degenerate continuum states with energy ε. The wave functions of the system atom plus field $\psi(t)$ has large components on the states spanning $\mathscr{E}_R$ and on the continuum states with energy close to the resonant one. If we had studied the evolution of $\psi(t)$ instead of $\mathbb{P}|\psi(t)\rangle$, the equations of motion replacing Eq. (3.3) would have been, in the case of a single resonance:

$$i\dot{a}_g = \Delta_g a_g + K a_r + \int \frac{d\varepsilon\, d\alpha}{\pi\hbar} J_{\varepsilon\alpha} a_{\varepsilon\alpha} \tag{7.1a}$$

$$i\dot{a}_r = (\delta + \Delta_r) a_r + K a_g + \int \frac{d\varepsilon\, d\alpha}{\pi\hbar} L_{\varepsilon\alpha} a_{\varepsilon\alpha} \tag{7.1b}$$

$$i\dot{a}_{\varepsilon\alpha} = \delta_\varepsilon a_{\varepsilon\alpha} + J_{\varepsilon\alpha} a_g + L_{\varepsilon\alpha} a_r \tag{7.1c}$$

$J_{\varepsilon\alpha}$ and $L_{\varepsilon\alpha}$, respectively, are the lowest order terms appearing in

$$\langle g|\langle N_0|R(E_0)|\varepsilon\alpha\rangle|N_0 - n\rangle \qquad \text{and} \qquad \langle r|\langle N_0 - p|R(E_0)|\varepsilon\alpha\rangle|N_0 - n\rangle$$

$$a_{\varepsilon\alpha} = \langle \varepsilon\alpha|\langle N_0 - n|\psi(r) \tag{7.2}$$

The states $|\varepsilon\alpha\rangle$ appearing in the integral of Eq. (6.1a) and (6.1b) are the states with energy E_ε close to resonance:

$$\hbar\delta = E_g + n\hbar\omega - E_\varepsilon \tag{7.3}$$

In studying the dynamics of the process we obtained $a_g(t)$ and $a_r(t)$ (Section IV). The probability for an electron to have appeared in states $|\varepsilon\alpha\rangle$ is $|a_{\varepsilon\alpha}|^2$, which can be obtained from a_g and a_r by

$$|a_{\varepsilon\alpha}|^2 = \int^t dt' \int^t dt'' \, e^{i\delta_\varepsilon(t'-t'')}[J_{\varepsilon\alpha}a_g(t') + L_{\varepsilon\alpha}a_r(t')][J^*_{\varepsilon\alpha}a^*_g(t'') + L^*_{\varepsilon\alpha}a^*_r(t'')] \tag{7.4}$$

Equation (7.4) may be used to describe the energy spectrum of the electrons (Knight, 1978, 1979; Armstrong and O'Neil, 1980; Coleman and Knight, 1981). If we are interested in the characteristics related to various α, but not in the energy of the electrons, then the probability for an electron to appear in state $|\varepsilon\alpha\rangle$ for any ε is

$$\int (d\varepsilon/\pi\hbar)|a_{\varepsilon\alpha}|^2 = \mathcal{N}_\alpha \tag{7.5}$$

From Eq. (7.4) we deduce

$$\mathcal{N}_\alpha = 2\int^t |J_{\varepsilon\alpha}a_g(t') + L_{\varepsilon\alpha}a_r(t')|^2 \, dt' \tag{7.6}$$

when $J_{\varepsilon\alpha}$ and $L_{\varepsilon\alpha}$ are supposed to be taken at the resonant value. The probability to reach any of the nearly resonant states $|\varepsilon\alpha\rangle$ is

$$P_\alpha = 2|J_{\varepsilon\alpha}a_g + L_{\varepsilon\alpha}a_r|^2 \tag{7.7}$$

This formalism may be used to study the spin polarization of electrons or the angular distribution of electrons (Leuchs et al., 1979a,b; Hansen et al., 1980; Dixit and Lambropoulos, 1981). In the latter case $J_{\varepsilon\alpha}$ and $L_{\varepsilon\alpha}$ are obtained by the methods used for nonresonant processes (Olsen et al., 1978). The angular distribution is studied in a plane perpendicular to the light beam, as a function of the angle θ with the polarization vector. For a non-resonant process, the angular distribution $P(\theta)$ is a polynomial nth order in $\cos^2\theta$, and reflects the structure of $J^2_{\varepsilon\alpha}$. For a resonant process the contribution $J_{\varepsilon\alpha}a_g$ in Eq. (7.7) may be negligible compared to the contribution $L_{\varepsilon\alpha}a_r$; in such a case the angular dependence of $P(\theta)$ is effectively given by $L^2_{\varepsilon\alpha}$ and thus corresponds to a polynomial in $\cos^2\theta$ of an order different from r.

VIII. CONCLUSION

We have presented here most of the methods proposed to study resonant multiphoton ionization. We have tried to emphasise the works which have led to, or could lead to, the interpretation of experiments. However, it is quite impossible to give an exhaustive survey of the literature. This is because, on the one hand, the large number of often overlapping papers devoted to the subject. On the other hand, an attempt to make a coherent and self contained description of the process leads unavoidably to some problems being treated in detail while others, no less fundamental, are just outlined.

If one looks at some of the formulae appearing above, resonant multiphoton ionization appears to be a complicated problem—to some extent, it is! However, even if some aspects of the formalism seem formidable, there has been a successful effort in the last few years to identify some simple features which characterize resonant processes. There have been only a few experiments in which all the parameters which enter a complete calculation have been measured, and it is remarkable that most of them have been successfully interpreted by theoretical calculations. This is the indication that the understanding of the process has reached a certain level, but it may also be that the process is not so complicated! Multiphoton ionization is a process occurring when the atom–light interaction is very strong and under conditions where many secondary processes have become negligible. From this point of view, resonant multiphoton ionization, in many cases, is a much simpler process than the atomic field interactions involving only discrete states. Although we have not avoided the tedious aspects of resonant multiphoton ionization, we have tried to emphasize the simplest models which allow a thorough discussion of the process and, in fact, are generally realistic enough to interpret the experimental data.

ACKNOWLEDGMENTS

It is a pleasure to thank Professors B.R. Judd and L. Armstrong, Jr. for their warm hospitality during my stay at The Johns Hopkins University where this paper was written. I am especially grateful to L. Armstrong for a critical reading of the manuscript. I am also indebted to Dr. Mireille Aymar for lively and illuminating discussions, which have been deepened our understanding of many aspects of this work. This work was supported by the National Science Foundation.

REFERENCES

Ackerhalt, J. R. (1978). *Phys. Rev. A* **17**, 293.
Ackerhalt, J. R., and Eberly, J. H. (1976). *Phys. Rev. A* **14**, 1705.
Ackerhalt, J. R., and Shore, B. W. (1977). *Phys. Rev. A* **16**, 277.

Agarwal, G. S. (1978). *Phys. Rev. A* **18**, 1490.
Agostini, P., Georges, A. T., Wheatley, S. E., Lambropoulos, P., and Levenson, M. D. (1978). *J. Phys. B* **11**, 1733.
Andrews, D. L. (1977). *J. Phys. B* **10**, L659.
Austin, E. J. (1979). *J. Phys. B* **12**, 4045.
Armstrong, L., Jr., and Baker, H. C. (1980). *J. Phys. B* **13**, 4727.
Armstrong, L., Jr., and O'Neil, S. V. (1980). *J. Phys. B* **13**, 1125.
Armstrong, L., Jr., Beers, B. L., and Feneuille, S. (1975). *Phys. Rev. A* **12**, 1903.
Armstrong, L., Jr., Lambropoulos, P., and Rahman, N. K. (1976). *Phys. Rev. Lett.* **36**, 952.
Aymar, M., and Crance, M. (1979). *J. Phys. B* **12**, L667.
Bakos, J. S. (1974). *Adv. Electron. Electron Phys.* **36**, 57.
Baravian, G., Godart, J., and Sultan, G. (1976). *Phys. Rev. A* **14**, 761.
Beers, B. L., and Armstrong, L., Jr. (1975). *Phys. Rev. A* **12**, 2447.
Brandi, H. S., and Davidovitch, L. (1979). *J. Phys. B* **12**, L615.
Chang, T. N., and Poe, R. T. (1976). *J. Phys. B* **9**, L311.
Chang, T. N., and Poe, R. T. (1977). *Phys. Rev. A* **16**, 606.
Chu, S. I., and Reinhardt, W. P. (1977). *Phys. Rev. Lett.* **39**, 1195.
Cohen-Tannoudji, C. (1967). *Cargese Phys.* **2**, 347.
Cohen-Tannoudji, C., and Haroche, S. (1969a). *J. Phys. (Orsay, Fr.)* **30**, 125.
Cohen-Tannoudji, C., and Haroche, S. (1969b). *J. Phys. (Orsay, Fr.)* **30**, 153.
Coleman, P. E., and Knight, P. L. (1981). *J. Phys. B* **14**, 2139.
Crance, M. (1978). *J. Phys., Lett. (Orsay, Fr.)* **39**, L68.
Crance, M. (1979). *J. Phys. B* **12**, 3655.
Crance, M. (1980). *J. Phys. B* **13**, 101.
Crance, M., and Aymar, M. (1979). *J. Phys. B* **12**, 3665.
Crance, M., and Aymar, M. (1980). *J. Phys. B* **13**, 4129.
Crance, M., and Feneuille, S. (1976). *J. Phys. (Orsay, Fr.)* **37**, L333.
Crance, M., and Feneuille, S. (1977). *Phys. Rev. A* **16**, 1587.
Dalgarno, A., and Lewis, J. T. (1955). *Proc. R. Soc. London, Ser. A* **233**, 70.
Debethune, J. L. (1972). *Nuovo Cimento B* **12**, 101.
Declemy, A., and Rachman, A., Jaouen, M., and Laplanche, G. (1981). *Phys. Rev. A* **23**, 1823.
Delone, N. B. (1975). *Sov. Phys. Usp. (Engl. Transl.)* **18**, 169.
de Meijere, J. L. F., and Eberly, J. H. (1978). *Phys. Rev. A* **17**, 1416.
Dixit, S. N. (1981). *J. Phys. B* **14**, L683.
Dixit, S. N., and Lambropoulos, P. (1978). *Phys. Rev. Lett.* **40**, 111.
Dixit, S. N., and Lambropoulos, P. (1979). *Phys. Rev. A* **19**, 1576.
Dixit, S. N., and Lambropoulos, P. (1980). *Phys. Rev. A* **21**, 168.
Dixit, S. N., and Lambropoulos, P. (1981). *Phys. Rev. Lett.* **46**, 1278.
Dixit, S. N., and Georges, A. T., Lambropoulos, P., and Zoller, P. (1980). *J. Phys. B* **13**, L157.
Edwards, M., and Armstrong, L., Jr. (1980). *J. Phys. B* **13**, L497.
Elgin, J. N. (1979). *Opt. Commun.* **30**, 150.
Faisal, F. H. M. (1973). *J. Phys. B* **6**, L89.
Faisal, F. H. M., and Moloney, J. V. (1981). *J. Phys. B* **14**, 3603.
Fano, U. (1961). *Phys. Rev.* **124**, 1866.
Fedorov, M. V. (1977). *J. Phys. B* **10**, 2573.
Fedorov, M. V., and Kazakov, A. E. (1977). *Opt. Commun.* **22**, 42.
Feneuille, S., and Armstrong, L., Jr. (1975). *J. Phys. (Orsay, Fr.)* **36**, L235.
Flank, Y, Laplanche, G., Jaouen, M., and Rachman, A. (1976). *J. Phys. B* **14**, L409.
Geltman, S. (1977). *J. Phys. B* **10**, 831.
Geltman, S. (1980). *J. Phys. B* **13**, 115.

Geltman, S., and Teague, M. R. (1974). *J. Phys. B* **7**, L22.

Georges, A. T., and Lambropoulos, P. (1977). *Phys. Rev. A* **15**, 727.

Georges, A. T., and Lambropoulos, P. (1978). *Phys. Rev. A* **18**, 587.

Georges, A. T., and Lambropoulos, P. (1979). *Phys. Rev. A* **20**, 991.

Georges, A. T., and Lambropoulos, P. (1980). *Adv. Electron Electron. Phys.* **54**, 191.

Goldberg, A., and Shore, B. W. (1978). *J. Phys. B* **11**, 3339.

Goldberger, M. L., and Watson, K. M. (1964). "Collision Theory," Chap. 8. Wiley, New York.

Gontier, Y., and Trahin, M. (1978). *J. Phys. B* **11**, L441.

Gontier, Y., and Trahin, M. (1979a). *J. Phys. B* **12**, 2123.

Gontier, Y., and Trahin, M. (1979b). *Phys. Rev. A* **19**, 264.

Gontier, Y., and Trahin, M. (1980). *J. Phys. B* **13**, 259.

Gontier, Y., Rahman, N. K., and Trahin, M. (1976). *Phys. Rev. A* **14**, 2109.

Gontier, Y., Rahman, N. K., and Trahin, M. (1981). *Phys. Rev. A* **24**, 3102.

Haan, S. L., and Geltman, S. (1982). *J. Phys. B* **15**, 1229.

Haken, H. (1970). *Handb. Phys.* **25**, 2.

Hansen, J. C., Duncanson, J. A., Jr., Chien, R. L., and Berry, R. S. (1980). *Phys. Rev. A* **21**, 222.

Heitler, W. (1954). "The Quantum Theory of Radiation." Oxford Univ. Press (Clarendon), London and New York.

Jaouen, M., Declemy, A., Rachman, A., and Laplanche, G. (1980). *J. Phys. B* **13**, L699.

Knight, P. L. (1978). *J. Phys. B* **11**, L511.

Knight, P. L. (1979). *J. Phys. B* **12**, L449.

Kovarskii, V. A., and Perelman, N. F. (1975). *Sov. Phys. JETP (Engl. Transl.)* **41**, 266.

Lambropoulos, P. (1973). *Phys. Rev. Lett.* **30**, 413.

Lambropoulos, P. (1974). *Phys. Rev. A* **9**, 1992.

Lambropoulos, P. (1976). *Adv. At. Mol. Phys.* **12**, 87–164.

Lambropoulos, P. (1980). *Appl. Opt.* **19**, 3926.

Lambropoulos, P., and Teague, M. R. (1976). *J. Phys. B* **9**, 587.

Lambropoulos, P., Doolen, G., and Rountree, P. (1975). *Phys. Rev. Lett.* **34**, 636.

Letokhov, V. S., and Mishin, V. I. (1979). *Opt. Commun.* **29**, 168.

Leuchs, G., Smith, S. J., Khawaja, E. E., and Walther, H. (1979a). *Opt. Commun.* **31**, 313.

Leuchs, G., Smith, S. J., and Walther, H. (1979b). *Springer Ser. Opt. Sci.* **21**, 255.

Lompré, L. A., Mainfray, G., Manus, C., and Thebault, J. (1978). *J. Phys., Lett (Orsay, Fr.)* **99**, 610.

Lompré, L. A., Mainfray, G., Mathieu, B., Watel, G., Aymar, M., and Crance, M. (1980). *J. Phys. B* **13**, 1799.

McClean, W. A., and Swain, S. (1978). *J. Phys. B* **11**, 1717.

McClean, W. A., and Swain, S. (1979a). *J. Phys. B* **12**, 2291.

McClean, W. A., and Swain, S. (1979b). *J. Phys. B* **12**, 723.

McGuire, E. J. (1981). *Phys. Rev. A* **23**, 186.

Mainfray, G., and Manus, C. (1980). *Appl. Opt.* **19**, 3934.

Messiah, A. (1965). "Quantum Mechanics." Wiley, New York.

Morellec, J., and Normand, D. (1979). *Multiphoton Processes, Proc. Int. Conf. Benodet*, 79.

Mostowski, J. (1976). *Phys. Lett. A* **56A**, 87.

Mower, L. (1966). *Phys. Rev.* **142**, 799.

Normand, D., and Morellec, J. (1981). *J. Phys. B* **14**, L401.

Olsen, T., Lambropoulos, P., Wheatley, S. E., and Rountree, S. P. (1978). *J. Phys. B* **11**, 4167.

Parzynski, R. (1979). *Opt. Commun.* **30**, 51.

Parzynski, R. (1980a). *Opt. Commun.* **34**, 361.

Parzynski, R. (1980b). *J. Phys. B* **13**, 469.

Parzynski, R. (1980c). *J. Phys. B* **13**, 475.

Pindzola, M. S., Chen, A. B., and Ritchie, B. (1981). *J. Phys. B.* **14**, 209.

Reiss, H. R. (1980). *Phys. Rev. A* **22**, 1786.
Ritchie, B. (1978). *Phys. Rev. A* **17**, 659.
Ritchie, B. (1979). *Phys. Rev. A* **20**, 1735.
Robinson, E. J. (1980). *J. Phys. B* **13**, 2243.
Sanchez, F. (1978). *J. Phys. (Orsay, Fr.)* **39**, L35.
Swain, S. (1979). *J. Phys. B* **12**, 3201.
Theodosiou, C. E., Armstrong, L., Jr., Crance, M., and Feneuille, S. (1979). *Phys. Rev. A* **19**, 766.
Wodkiewicz, K. (1979). *J. Math. Phys. (N. Y.)* **20**, 45.
Wong, N. C., and Eberly, J. H. (1977). *Opt. Lett.* **1**, 211.
Yeh, J. J., and Eberly, J. H. (1981). *Phys. Rev. A* **24**, 888.
Zoller, P. (1979). *Phys. Rev. A* **19**, 1151.
Zoller, P., and Lambropoulos, P. (1979). *J. Phys. B* **12**, L547.
Zoller, P., and Lambropoulos, P. (1980). *J. Phys. B* **13**, 69.

5

Angular Distribution of Photoelectrons and Light Polarization Effects in Multiphoton Ionization of Atoms

*G. LEUCHS**

Sektion Physik
Universität München
Garching, Federal Republic of Germany

H. WALTHER

Sektion Physik
Universität München and
Max-Planck-Institut für Quantenoptik
Garching, Federal Republic of Germany

I. INTRODUCTION

Multiphoton ionization processes are usually characterized by their total cross sections. Measurements of the total cross section are straightforward but not necessarily easy. Additional information can be obtained by studying the kinetic energy, the angular distribution, and the spin polarization of the

* Present address: Joint Institute for Laboratory Astrophysics, University of Colorado, and National Bureau of Standards, Boulder, Colorado.

photoelectrons. Also the ejected electron may leave the ion in a state with spatial anisotropy, where not all magnetic substates are equally populated.

The results depend on different parameters of the photoionization experiment, namely polarization, wavelength, and bandwidth of the incident laser radiation, as well as its intensity, which in most experiments is not constant in time. The cross section for multiphoton ionization may increase by many orders of magnitude if the atoms are irradiated by laser light which is in resonance with intermediate states of the atom.

In this chapter we shall focus on the angular distribution of photoelectrons in resonant and nonresonant multiphoton ionization. We shall also discuss the related effect of laser polarization on the angle-integrated cross sections.

The interest in differential photoionization cross sections has existed for a long time, starting soon after single-photon ionization, then known as the photoeffect, had been observed. The first refined measurements of the electron angular distribution in one-photon ionization have been performed by Bothe (1924) when ionizing molecules with x rays. For such high-energy photons the linear momentum of the photon is not negligible with respect to the momentum of the bound electron. This results in the famous distortion of the photoelectron angular distribution due to momentum transfer (Sommerfeld and Schur, 1930). The first measurements of the angular distribution of photoelectrons in single-photon ionization of groundstate atoms and of molecules with photon energies less than 10 eV have been performed by Berkowitz and Ehrhardt (1966). At these photon energies the photon momentum transfer can be neglected.

In multiphoton ionization, angular distributions of electrons were first observed by Edelstein et al. (1974). Since then measurements of the angular distributions have been used to study various features of the atomic electronic structure, e.g., the phase differences among continuum state wave functions (Strand et al., 1978; Kaminski et al., 1980; Siegel et al., 1982), characteristics of autoionizing states (Feldmann and Welge, 1982), quantum interference effects (Leuchs et al., 1979a), and perturbations of intermediate levels by configuration mixing (Leuchs et al., 1983). Recently, the influence of the ac-Stark effect on angular distributions has also been discussed theoretically (Dixit and Lambropoulos, 1981) and experimentally (Leuchs et al., 1980, 1982; Ohnesorge et al., 1983). Apart from the fundamental theoretical interest in photoelectron angular distribution data, it is possible to obtain detailed information about photoionization and recombination processes involving excited atomic states. These processes are of considerable importance e.g., for astrophysics, but are little studied experimentally to date.

The angular distributions depend sensitively on the intermediate state quantum numbers. Therefore relaxation processes between magnetic substates induced for example by collisions can be studied (Tully *et al.*, 1968; Lambropoulos and Berry, 1973). Also radio-frequency transitions between closely spaced excited states could be detected by the corresponding change in the angular distribution (Leuchs *et al.*, 1979b).

In the following sections we first briefly review the theoretical work on photoelectron angular distributions and then discuss several experiments.

II. ANGULAR DISTRIBUTION OF PHOTOELECTRONS IN MULTIPHOTON IONIZATION

For photons having an energy not higher than a few electron volts, the transfer of the photon linear momentum can be neglected. In this approximation the angular distribution of photoelectrons in one-photon ionization of ground-state atoms given by the differential cross section for photoionization can be written as (Wentzel, 1927)

$$\frac{d\sigma^{(1)}}{d\Omega} = \frac{\sigma_{\text{tot}}}{4\pi} \left[1 + \beta_2 P_2(\cos \Theta) \right] \tag{2.1}$$

where Θ is the angle between the direction of the outgoing electron and the quantization axis, which is most conveniently taken to be parallel to the direction of polarization of the ionizing light; P_2 is the second Legendre polynomial; and σ_{tot} is the angle integrated (total) photoionization cross section. This equation for the angular distribution reflects the dipole character of the interaction between the photon and the atom. In experiments on one-photon ionization, the anisotropy parameter β_2 is measured, e.g., as a function of the wavelength of the ionizing light (Samson, 1982).

Equation (2.1) is correct only if all magnetic substates of the initial state of the atom are equally populated, i.e., if the initial state is isotropic. For resonant or nonresonant ionization with two or more photons, higher-order Legendre polynomials have to be included.

In the case of multiphoton ionization of an atom in an initially isotropic state, the spatial anisotropy of the atom may be increased with each photon absorbed. The general formula describing the resulting differential cross section for the emission of a photoelectron is

$$\frac{d\sigma^{(N)}}{d\Omega} = \frac{\sigma_{\text{tot}}^{(N)}}{4\pi} \sum_{j=0}^{N} \beta_{2j} P_{2j}(\cos \Theta) \tag{2.2}$$

where N is the number of photons absorbed by each atom and $\sigma_{\text{tot}}^{(N)}$ is the so-called generalized total cross section for N-photon ionization as defined, e.g., by Lambropoulos (1976). Equation (2.2) can be verified on the basis of a theorem by Yang (1948), which was developed for the analysis of angular correlation measurements in nuclear physics.

A. Theory

Electron angular distributions in multiphoton ionization have first been calculated by Zernik (1964) for the special case of two-photon ionization of the 2s metastable state of hydrogen. Tully *et al.* (1968) and Lambropoulos and Berry (1973) pointed out that angular distribution measurements in resonant two-photon ionization may be used to observe relaxation of coherence in the intermediate state. Several theoretical publications dealt with the problem of separating the contribution of angular momentum quantum numbers of initial and intermediate atomic states and of the ionizing photons, on the one hand, and of radial matrix elements, on the other hand (Cooper and Zare, 1969; Lambropoulos, 1972a,b; Mizuno, 1973; Arnous *et al.*, 1973; Jacobs, 1973; Gontier *et al.*, 1975; Lambropoulos and Teague, 1976; Teague and Lambropoulos, 1976). An extensive theoretical treatment for resonant multiphoton ionization, including effects of high laser intensity, has been presented by Dixit and Lambropoulos (1983). The earlier work has been discussed in the review article by Lambropoulos (1976). Closely related are calculations of photoelectron angular distributions in one-photon ionization out of aligned or polarized atomic states (Jacobs, 1972; Kollath, 1980; Klar, 1980; Klar and Kleinpoppen, 1982).

1. Basic Formulas for Photoelectron Angular Distributions

In the case of resonant multiphoton ionization, the resonant intermediate states in general have spatial anisotropy. Therefore the calculation of the electron angular distribution in photoionization out of a state with a well-defined magnetic quantum number m will be sketched briefly. The result can also be used to give the general formula for the angular distribution of photoelectrons in nonresonant multiphoton ionization. The differential photoionization cross section is obtained by evaluating the electric dipole matrix element for the transition from the initial bound state to the continuum. The effects of spin–orbit and hyperfine coupling have been discussed in several theoretical papers (see, e.g., Lambropoulos and Teague, 1976; Teague and Lambropoulos, 1976; Kollath, 1980; and references therein), but, for the sake of simplicity, will be ignored here. The wave function of the bound state is characterized by the principal, angular momentum, and magnetic quantum numbers. For a potential with inversion symmetry it factorizes into a radial and an angular part:

$$\psi_{nlm}(\mathbf{r}) = R_{nl}(r)Y_{lm}(\theta, \phi) \qquad (2.3)$$

The continuum state is taken as a superposition of an incoming spherical and an outgoing plane wave, and can be expanded in Legendre polynomials (Cooper and Zare, 1969; Lambropoulos, 1976), which in turn can be expanded in products of spherical harmonics:

$$\psi_{\mathbf{k}}(\mathbf{r}) = \sum_{l'=0}^{\infty} i^{l'} \exp(i\delta_{l'}) 4\pi G_{kl'}(r) \sum_{m=-l'}^{l'} Y_{l'm}^*(\Theta, \Phi) Y_{l'm}(\theta, \phi) \qquad (2.4)$$

The arguments of the spherical harmonics contain the angles Θ, Φ and θ, ϕ, which describe the direction of the electron wave vector $\mathbf{k}$ and the radius vector $\mathbf{r}$, respectively. The angular distribution of the photoelectron is then proportional to the absolute value of the square of the electric dipole matrix element

$$\frac{d\sigma}{d\Omega} \approx |\langle \mathbf{k} | \varepsilon \cdot \mathbf{r} | nlm \rangle|^2 \qquad (2.5)$$

where ε is the polarization of the ionizing light. For light polarized linearly in the z direction, we have $\varepsilon = (0, 0, 1)$. The corresponding selection rules are $\Delta l = \pm 1$. With these selection rules only two terms in the infinite sum in Eq. (2.4) that describes the continuum state have to be considered.

Then $d\sigma/d\Omega$ is the absolute square of the amplitudes of two outgoing partial waves $l' = l + 1$ and $l' = l - 1$. The absolute square contains an interference term that depends on the difference of the corresponding scattering phases $\delta_{l+1} - \delta_{l-1}$. The information about the scattering phase is available only in the angular distribution (Lambropoulos, 1976). When measuring the total cross section σ_{tot}, the angular distribution has to be integrated over Θ and Φ. In this case, the interference term vanishes owing to the orthogonality of the spherical harmonics.

For the purpose of deriving Eq. (2.1) for one-photon ionization and the corresponding Eq. (2.2) for N-photon ionization, the problem is simplified here by neglecting the $(l - 1)$-partial wave. Of course for almost any special case this simplification would yield wrong numbers. It is, however, useful for the purpose of determining the highest order of anisotropy to be expected for the angular distribution. Using this simplification the angular distribution is

$$\frac{d\sigma}{d\Omega} \approx |Y_{l+1,m}(\Theta, \Phi)|^2 \frac{(l+1)^2 - m^2}{(2l+1)(2l+3)} A_{l+1}^2 \qquad (2.6)$$

Here the electric dipole operator $\varepsilon \cdot \mathbf{r}$ is written as $r \cdot \cos \theta$ for linearly polarized light. The radial integral, represented by A_{l+1}, is

$$A_{l+1} = 4\pi \int_0^\infty dr \, r^3 G_{k,l+1}(r) R_{n,l}(r)$$

The angular part of the electric dipole matrix element has already been evaluated in Eq. (2.6):

$$\int_0^\pi d\theta \sin\theta \cos\theta \int_0^{2\pi} d\phi\, Y^*_{l+1,m}(\theta,\phi) Y_{l,m}(\theta,\phi) = \left(\frac{(l+1)^2 - m^2}{(2l+1)(2l+3)}\right)^{1/2} \quad (2.7)$$

In the case where all magnetic sublevels of the bound state are equally populated, $d\sigma/d\Omega$ of Eq. (2.6) has to be summed over m. Using the relation

$$\sum_{m=-l}^{l} m^{2M} |Y_{l,m}(\Theta,\Phi)|^2 = \sum_{j=0}^{M} a_j \sin^{2j}(\Theta) \quad (2.8)$$

one obtains, for the m-averaged angular distribution [Eq. (2.6)], a constant plus a $\sin^2\Theta$ term, and thus the same general form as Eq. (2.1). The coefficients a_j in Eq. (2.8) can be calculated by differentiating the addition theorem for spherical harmonics with respect to Φ (F. Diedrich, private communication, see also Cooper and Zare, 1969).

The physical interpretation is that the original system consists of the spatially isotropic atom and the incoming photon. Since the interaction of the photon and the atom has dipole character, the total system has dipole anisotropy, which is also displayed in the photoelectron angular distribution. Considering the case of N-photon absorption, the N dipole interactions lead to an angular distribution containing even powers of m to $2N$. The angular distribution is again averaged over the m sublevels of the initial state. As a result one obtains Eq. (2.2).

Again the physical interpretation is that in each absorption step the spatial anisotropy of the initially isotropic atoms can be increased because of the dipole character of the photon–atom interaction. Since the physical interpretation is not affected by the coupling scheme, Eq. (2.2) holds also in the case of spin–orbit and hyperfine coupling.

The coefficient β_{2N} of the most-anisotropic Legendre polynomial in Eq. (2.2) is zero if the anisotropy of the atom is not increased in one of the absorption steps. This can take place only in resonant multiphoton ionization when, e.g., two successively excited states have the same number of m sublevels and the laser light is linearly polarized.

Another factor complicating the calculation of the photoelectron angular distribution is that the remaining ion may carry some anisotropy. This, however, does not apply to the ionization of alkali atoms with photons having an energy of a few electron volts. The ground state of singly ionized alkali atoms is a $^1S_0$ state and the excited states of the ions lie so high that they are not accessible by photoionization with visible laser light. If several lasers that do not all have the same direction of linear polarization are used, the angular distribution also depends on Φ. The maximum possible anisotropy, however, is still determined by the number of photons absorbed.

B. Experiments

The angular distribution of photoelectrons in multiphoton ionization was first observed by Edelstein *et al.* (1974) when ionizing titanium atoms in a resonant two-photon process. In subsequent experiments on sodium, cesium, strontium, neon, and barium, the influence of resonant intermediate states on the angular distribution was studied. Experiments observing photo-electron angular distributions in nonresonant multiphoton ionization were performed on Xe and Na.

1. Two- and Three-Photon Ionization of Sodium Atoms via Resonant Intermediate States

In this section two selected experiments will be presented. The results are used to discuss qualitatively the simple physical pictures introduced in Section II.A.1. In an experiment by Hellmuth *et al.* (1981), a nitrogen-laser–pumped-dye laser was used to excite sodium atoms in a collimated, thermal beam to the 3p ^{2}P state. Part of the nitrogen laser radiation, which was not used to pump the dye laser, was aligned to be collinear with the dye laser beam and was also focused onto the atomic beam, ionizing the sodium atoms in the 3 ^{2}P state. The ejected photoelectrons have a kinetic energy of about 0.6 eV (Fig. 1). The durations of the dye and the nitrogen laser pulses were 4 and 6 ns, respectively. The pulses were adjusted in time so that they inter-acted simultaneously with the atoms. The earth's magnetic field was com-pensated for at the interaction region, which was also enclosed between

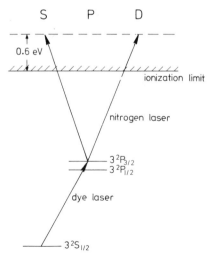

Fig. 1. Part of the energy level scheme of sodium showing the transitions used for two-photon ionization.

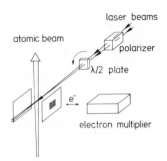

Fig. 2. Sketch of the interaction region. Two half-wave plates are used to rotate the linear polarization directions in the plane, perpendicular to the propagation direction of the light, one for each laser beam. The figure shows only one of the plates.

parallel electrically grounded metal plates serving as a shield against external electric fields. The emitted photoelectrons passed a wire-mesh-covered aperture in one of the plates and were accelerated through 200 V onto the cathode of a 17-stage electron multiplier (Fig. 2). The angular distribution of the photoelectrons was measured in the plane perpendicular to the propagation direction of the laser beams. For this purpose, the positions of aperture and detector remained fixed and the linear polarizations of both laser beams were rotated, using half-wave plates.

Figure 3 shows polar diagrams of the number of photoelectrons measured as a function of the angle Θ between the direction of laser polarization and the emission direction of the photoelectron. The polarization directions

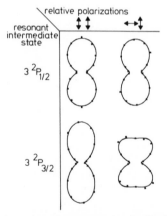

Fig. 3. Polar diagrams of the photoelectron angular distributions measured in two-photon ionization of sodium (points). The result of the least-squares fit of Eq. (2.2) to the data is also shown (solid line). The arrows on the top of the figure indicate the polarization directions of the dye (left arrow) and the nitrogen (right arrow) lasers. [From Hellmuth *et al.* (1981).]

of both lasers have been adjusted parallel and perpendicular to each other. In the latter case the angle Θ in the polar diagram was taken with respect to the vertical polarization, which is the polarization direction of the ionizing laser beam. The wavelength of the dye laser was tuned into resonance with either the 3p $^2P_{3/2}$ or the $^2P_{1/2}$ state. It is apparent that, in the case of ionization via the $^2P_{1/2}$ state, the shape of the angular distribution does not change when the polarization directions for both laser beams are adjusted from parallel to perpendicular to each other. However, a clear polarization dependence is observed when ionizing through the $^2P_{3/2}$ state. This demonstrates the effect of the spatial anisotropy of the intermediate state. Exciting sodium atoms from the $3\,^2S_{1/2}$ ground state to the $3\,^2P_{1/2}$ state with linearly polarized light ($\Delta m = 0$ selection rule) all magnetic substates of the $3\,^2P_{1/2}$ state are equally populated, whereas in the case of the $3\,^2P_{3/2}$ state only half the magnetic substates are excited (Fig. 4).

As expected for two-photon ionization, the least-squares fit of Eq. (2.2) to the data gave nonzero values only for β_0, β_2, and β_4. When ionizing via the $^2P_{1/2}$ state even β_4 was zero within the error bars being a result of the isotropic population of the m substates of the intermediate state as mentioned above.

In a similar experiment two dye lasers and a Nd:YAG laser were used to ionize sodium atoms of a thermal beam via the $3\,^2P$ and $20\,^2D$ resonant intermediate states as shown in Fig. 5 (Leuchs, 1983, and references therein). The two dye lasers were pumped by the second and third harmonic of the Nd:YAG laser, and the pulse duration was ~ 6 ns. The Nd:YAG laser pulse at a wavelength of 1.064 μm, used to ionize sodium atoms in the $20\,^2D$-state,

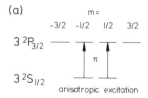

(a)

$3\,^2P_{3/2}$

$3\,^2S_{1/2}$

anisotropic excitation

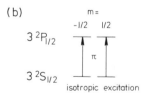

(b)

$3\,^2P_{1/2}$

$3\,^2S_{1/2}$

isotropic excitation

Fig. 4. Effect of the intermediate state angular momentum quantum number on the spatial anisotropy transferred to the atom.

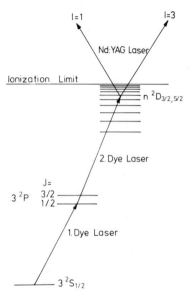

Fig. 5. Part of the level scheme of sodium showing the transitions used for three-photon ionization.

was delayed by ~ 30 ns with respect to the dye laser pulses. Therefore, the high intensity of the ionizing laser radiation does not perturb the atomic structure during the excitation of the $20\,^2\mathrm{D}$ state. This ensures a pure stepwise ionization.

The three laser beams were propagating nearly collinearly, the maximum deviation being less than $2°$. The angular distributions were recorded in a plane perpendicular to the propagation direction of the laser beams by inserting one achromatic half-wave plate in the two dye laser beams and one low-order half-wave plate in the Nd:YAG-laser beam and rotating them simultaneously with stepping motors. Figure 6 shows polar diagrams of photoelectron angular distributions recorded for various combinations of polarizations and for two different first intermediate states $3\,^2\mathrm{P}_{1/2}$ and $3\,^2\mathrm{P}_{3/2}$. On the top of each column in Fig. 6, the three arrows indicate the polarizations of the lasers for the first, the second, and the final step. In the case where circularly polarized light was also used, only the linear polarization of the ionizing laser beam was rotated using a half-wave plate. The data demonstrate the sensitivity of the photoelectron angular distributions on the m-quantum number of the intermediate state.

The evaluation of these angular distributions shows that for Na $n\,^2\mathrm{D}$ states the $(\Delta l = -1)$ partial wave has a negligible amplitude as compared to the $(\Delta l = 1)$ partial wave (Leuchs and Smith, 1981). This result can also

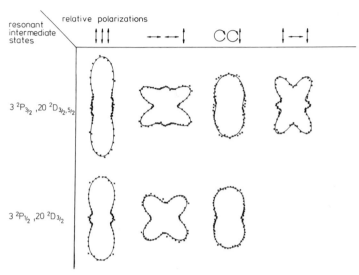

Fig. 6. Polar diagrams of photoelectron angular distributions in three-photon ionization of sodium. The arrows on the top of the figure indicate the polarization directions of dye lasers 1 and 2 and the ionizing Nd:YAG laser. [From Leuchs (1983).]

be obtained in a measurement of the total cross section, however, only if the intermediate state is fully polarized (Smith *et al.*, 1980; see also discussion in Section II.B.2).

The angular distributions obtained with the first laser tuned to the $3\,^2P_{3/2}$ state, shown in the upper row of polar diagrams in Fig. 6, are sensitive to the duration and time delay between the laser pulses. This is a result of the unresolved hyperfine and fine structure of the $3\,^2P_{3/2}$ and the $20\,^2D$ states, leading to quantum interference effects (Leuchs *et al.*, 1979a). If there is no time delay between the laser pulses as in the experiments described here, the angular distributions can be smeared out because of the quantum interference effects. This effect—discussed further in Section II.B.5—is reduced when using shorter pulses, obtainable from nitrogen-laser-pumped dye lasers (Leuchs *et al.*, 1979b; Leuchs and Smith, 1981), as can be seen in a comparison of Figs. 6 and 7.

A final remark will be made regarding the similarity of the second and fourth angular distribution displayed in the upper row of polar diagrams in Fig. 6. The main difference seems to be a rotation by 90°. But the corresponding laser polarizations indicated at the top of the figure are not the result of a simple rotation. A study of the selection rules, however, shows that, neglecting the smearing effects mentioned above, the angular distributions should just be the same, apart from the rotation by 90°.

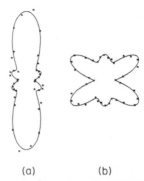

(a)   (b)

Fig. 7. Photoelectron angular distributions similar to the ones shown in Fig. 6, only the two dye-laser pulse durations were shorter, resulting in a more pronounced anisotropy. The dye lasers were tuned to the $3\,^2S_{1/2} \rightarrow 3\,^2P_{3/2} \rightarrow 20\,^2D$ transitions. (a) All linear polarizations directions were parallel to each other and (b) the ionizing laser pulse was polarized perpendicular to the two dye-laser pulses. [Reprinted with permission from Leuchs and Smith (1983). Copyright 1983 by the Institute of Physics.]

2. Bound-Free Matrix Elements and Scattering Phases

In Section II.A.1 it became apparent that photoelectron angular distributions contain information about the scattering phases of the outgoing partial waves that is not available in total cross section measurements (Lambropoulos, 1976). Additional information on the scattering phases can be obtained from the measurement of the angular distribution of the electron spin polarization (Kollath, 1980). The cosine of the difference of the scattering phases of the two outgoing partial waves have been measured in resonant two-photon ionization of sodium atoms via the $3\,^2P_{1/2,\,3/2}$ excited states (Duncanson et al., 1976; Strand et al., 1978; Hansen et al., 1980). These experiments were performed on sodium atoms in a thermal beam irradiated at right angles by simultaneous dye- and nitrogen-laser pulses.

Measurements of angular distributions of photoelectrons and their spins in resonant two-photon ionization of cesium atoms via the $7\,^2P$ states by Kaminski et al. (1980) lead to an experimental value for the difference of scattering phases for the s and d partial waves. In this experiment, however, there was a problem with the interpretation of the results since a flash-lamp-pumped dye laser having a pulse duration of 400 ns was used. For such long pulses, the theoretical calculation, which finally allows one to extract the phase information, should neither be performed in the fine structure nor in the hyperfine structure, but in an intermediate coupling scheme. Under such circumstances the outcome of the experiment will also strongly depend on the temporal structure of the dye laser pulse. The influence of the duration of the laser pulses on angular distributions will be discussed in Section II.B.5 on quantum interference effects.

Recently, Siegel *et al.* (1982, 1983) used continuous-wave dye lasers for resonant two-photon ionization of neon atoms in the 3s $^3P_2$ metastable state. The 3p $^3D_3$ resonant intermediate state was optically aligned. From the angular distributions of the photoelectrons, the scattering phases and ratios of radial matrix elements were deduced.

Total photoionization cross-section measurements provide no information about the scattering phases of the partial waves. However, partial cross sections corresponding to the absolute square of the radial matrix elements for different partial waves can be obtained in total cross-section measurements (Lambropoulos, 1972a; Klarsfeld and Maquet, 1972; Lambropoulos, 1973; Mizuno, 1973). For references to experimental work see Lambropoulos (1976) and Dixit (1981). A direct way of obtaining the partial cross sections for different partial waves has been demonstrated by Duong *et al.* (1978). Using a circularly polarized continuous-wave dye laser sodium atoms of a thermal beam were optically pumped to the $F = 3$, $M_F = 3$ magnetic hyperfine level of the $3\,^2P_{3/2}$, state. The excited sodium atoms were ionized with pulsed *uv*-laser radiation having either the same or the opposite circular polarization. As a result, the atoms are ionized via the $l \to l + 1$ channel alone or via both $l \to l \pm 1$ channels, allowing for an independent measurement of the partial cross sections. In this experiment, however, the partial cross sections were measured only relative to one another. For an absolute determination of the cross sections the atom number density has to be measured. A way to avoid this has been pursued by Smith *et al.* (1980) by working at a level of the ionizing laser intensity, where the ionization signal saturates with intensity (Arlanbekov *et al.*, 1975). Combining the above-mentioned technique of optical pumping with saturated ionization, Smith *et al.* (1980) obtained absolute values for the partial ionization cross sections for photoionization out of the $4\,^2D$ state of sodium.

Although absolute values of radial matrix elements can be obtained from the angle-integrated ionization signal by making use of light polarization effects, it can nevertheless be advantageous to measure the photoelectron angular distribution in addition (Amusia *et al.*, 1972). The interference term in the angular distribution—discussed in Section II.A.1—contains the product of the radial matrix elements for both partial waves. In the case where the amplitude of one partial wave is weak and the other is strong, the weak partial wave shows up more pronounced in the angular distribution than in the partial cross sections that are absolute squares of radial matrix elements.

The discussion so far was concentrating on multiphoton ionization via resonant intermediate states. In the case of nonresonant multiphoton ionization light polarization effects can also be of large influence (Klarsfeld and Maquet, 1972; Lambropoulos and Teague, 1976; Teague and Lambropoulos, 1976). The interpretation is, however, more complicated since, in general, a larger number of partial waves is contributing, resulting from the increase of the number of possible ionization channels, and since the theoretical calculation involves infinite summations over intermediate states (Lambropoulos, 1976). For a further discussion see Section II.B.7.

3. Perturbation by Configuration Mixing

In experiments on alkali atoms like the ones on sodium described in Section II.B.1, the shape of the photoelectron angular distribution is expected to depend only on the angular momentum quantum numbers and not on the principal quantum number. This is not the case when atoms with more than one valence electron are studied, where doubly excited levels exist below the first ionization limit. The configurations with only one excited electron mix strongly with those having two excited valence electrons whenever their energies are close to each other. This is, e.g., the case in barium for the $6snd\ ^{1,3}D_2$ series, which in the vicinity of $n = 27$ is perturbed by the $5d7d\ ^1D_2$ state. When ionizing such highly excited states of barium with the fundamental radiation of a Nd:YAG laser, an energy is reached in the continuum which is higher than the ionization limit of the doubly excited series $5dnl$ (Fig. 8). As a result, the remaining ion may be left either in the $6s\ ^2S_{1/2}$ ground or the $5d\ ^2D_{3/2,\,5/2}$ excited state. An ion in the excited state, however, may carry some spatial anisotropy. Quite different photoelectron angular distributions are therefore expected for photoionization leaving the ion either in the 6s or in the 5d state.

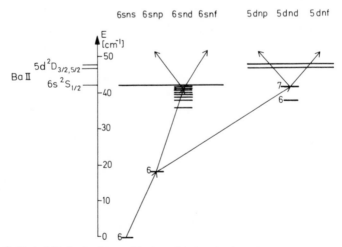

Fig. 8. Part of the level scheme of barium showing the 6s and 5d continua, both of which can be reached from highly excited $^{1,3}D_2$ states by using Nd: YAG-laser radiation.

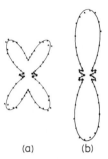

(a)   (b)

Fig. 9. Photoelectron angular distributions in resonant three-photon ionization of barium via (a) the 6s19d ^{3}D and (b) 6s19d ^{1}D state. Only electrons ionizing into the 6s continuum were measured. [From Matthias *et al.* (1983).]

In a recent experiment, two dye lasers pumped by the second and third harmonic of a Nd:YAG laser were used to excite barium atoms via the $6s6p^1P_1$ state to states of the 6s*nd* series. The subsequent ionization by the Nd:YAG-laser radiation at the fundamental frequency was time delayed to avoid light shifts during the excitation (Leuchs *et al.*, 1983; Matthias *et al.*, 1983; see also Leuchs, 1983). Electrons resulting from ionization into the 6s or 5d state of the remaining ion could be distinguished by time-of-flight analysis.

The experiment showed that at $n = 19$, where singlet–triplet mixing owing to the perturber is rather weak, the angular distributions are remarkably different, when the laser frequency is tuned into resonance with the 6s19d $^1D_2$ or the 6s19d $^3D_2$ intermediate state (Fig. 9). (See also Olsen *et al.*, 1978.) Consequently, the angular distributions are sensitive to singlet–triplet mixing produced, e.g., by a perturbing state. It has also been seen that the shape of the angular distribution is very different for electrons originating from the 6s or 5d continuum, the latter ones being more isotropic since the ion also carries some anisotropy.

Figure 10 shows a series of angular distributions of 6s continuum electrons for which the dye lasers were tuned into resonance with different members of the 6s*nd* $^1D_2$ series. The resonancelike increase of state mixing in the vicinity of the perturbing 5d7d $^1D_2$ state is obvious. The analysis of these data (Matthias *et al.*, 1983) has shown that the detailed information obtained from angular distribution measurements can be used to improve the parameters of multichannel quantum defect calculations.

4. Autoionizing States

In atoms with two or more valence electrons there exist low lying configurations with more than one electron excited. The total energies of these

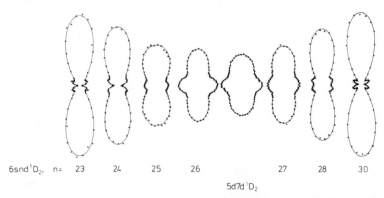

6snd^{1}D$_2$, n=   23   24   25   26   27   28   30

5d7d ^{1}D$_2$

Fig. 10. Photoelectron angular distributions similar to the one in Fig. 9b except that the principal quantum number of the intermediate state was varied from $n = 23$ to $n = 30$. The angular distribution via the perturbing 5d7d ^{1}D$_2$ intermediate state is also shown. [From Leuchs *et al.* (1983).]

levels lie, in general, above the ionization limit of a single valence electron. In this case there are two decay mechanisms for the isolated atom, emission of radiation, and ionization via a coupling to the continuum above the first ionization limit. The latter process is called autoionization.

Autoionizing states which can be excited starting from the ground state by one-photon transitions have been studied extensively (see Garton, 1978; and references therein). Multiphoton transitions allow one to reach auto-ionizing states with the same parity as the ground state and with higher angular momentum than in direct excitation. In an experiment by Cooke and Gallagher (1978), two autoionizing states of barium which overlap within in their line width have been studied independently in a three-step resonant laser excitation using light polarization effects.

First electron angular distributions in multiphoton ionization via auto-ionizing resonances have been observed by Feldmann and Welge (1982). In this work on strontium atoms autoionizing resonances have been found in three-photon transitions, where the change in the angular momentum quantum number was three. Part of the ionization signal was also due to four-photon ionization where, one additional photon was absorbed by the strontium atoms in the autoionizing state. Using electric retarding fields, the two electron signals were discriminated and the electron angular distributions have been measured separately. Apart from a small asymmetry, not yet understood, the angular distributions comply with Eq. (2.2). First calculations for this system including laser intensity effects have been published by Kim and Lambropoulos (1982).

In a recent experiment by Sandner *et al.* (1983), the 6pns autoionizing resonances of barium have been studied in three-photon ionization for $n = 9$ to $n = 20$. The photoelectron kinetic energies and angular distributions were measured yielding results for the branching ratios of the autoionizing decay into the different states of the remaining ion.

5. Quantum Interference Effects

A quantum-mechanical system prepared in a coherent superposition of energy eigenstates may exhibit some oscillatory time evolution if the eigenstates are not degenerate. Examples are the perturbed angular correlation of γ quanta emitted in a cascade decay of a nucleus (Abragam and Pound, 1953) and quantum beats observed in the decay of the fluorescence of an atom following pulsed laser excitation (Alexandrov, 1964; Dodd et al., 1964; Haroche, 1976). Apart from the detection of the fluorescence, quantum beats can be observed in the absorption to a higher bound state (Ducas et al., 1975), in field ionization (Leuchs and Walther, 1977), in four-wave mixing (Lange and Mlynek, 1978), and in photoionization (Leuchs et al., 1979a).

The oscillatory response of the atom can be understood in analogy to Young's double-slit experiment. There are two (or more) different channels for the atom when the transition from the initial to the final state is performed via two (or more) closely spaced energy eigenstates. The indistinguishability of these channels leads to the quantum interference effects.

A necessary condition for the observability of quantum beats is that the initially excited coherent superposition of energy eigenstates corresponds to a spatially anisotropic excitation of the atom. This spatial anisotropy will evolve in time and may be observed as quantum beats (Ducas et al., 1975). If, however, the coherent superposition of states corresponds to a spatially isotropic atomic excitation, no quantum beats will be observed.

In this section the observation of quantum beats in the angular distribution of photoelectrons will be discussed. Theoretically the problem was addressed by Zygan-Maus and Wolter (1978), Georges and Lambropoulos (1978), and Knight (1979).

The atomic system considered here is sodium and the closely spaced energy eigenstates are the hyperfine structure states of the $3\,^2P_{3/2}$ level. Since the spin of the ^{23}Na nucleus is $I = \frac{3}{2}$, the total angular momentum quantum number may have four different values $F = 0, 1, 2, 3$. Photoelectron angular distributions observed for this system have already been discussed in Section II.B.1 in Fig. 3. There the hyperfine structure of the $3\,^2P_{3/2}$ state was not mentioned since the laser pulses were shorter than the shortest hyperfine period. It has, however, been demonstrated that using linearly polarized light all magnetic substates of the $3\,^2P_{1/2}$ are populated equally, resulting in a spatially isotropic excited atomic state. Consequently, it is not possible to observe quantum beats for the hyperfine states, although a coherent superposition of the $F = 1$ and $F = 2$ states of the $3\,^2P_{1/2}$ is excited.

For the observation of the quantum beats (Leuchs *et al.*, 1979a; Hellmuth *et al.*, 1981), the dye-laser radiation was tuned to excite the $3\,^2P_{3/2}$ state. The pulse duration was 4 ns, short enough so that the Fourier-transformed bandwidth overlaps all hyperfine structure states. The ionizing nitrogen-laser pulse of 6-ns duration was then delayed using a variable optical delay line, probing the state of the atomic evolution at different times. The angular distributions of the resulting photoelectrons were measured for various delays and for the polarization directions of both lasers being parallel and perpendicular to each other (Fig. 11). The experimental data shown as the crosses in the polar diagrams were fitted using Eq. (2.2) with $N = 2$. The coefficient β_4 obtained in the least-squares fit is also plotted as a function of the delay time (Fig. 11). The periodic behavior is obvious. A Fourier transformation of $\beta_4(t)$ yields frequencies in good agreement with the known hyperfine structure (Leuchs *et al.*, 1979a).

The periodic variation of $\beta_4(t)$ for parallel and perpendicular polarizations displays the same frequencies, but both signals are 180° out of phase

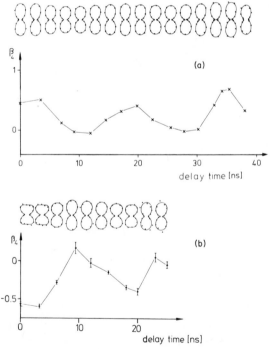

Fig. 11. Angular distributions in two-photon ionization of sodium via the $3\,^2P_{3/2}$ resonant intermediate state. The ionizing laser pulse was time delayed with respect to the pulsed excitation of the $3\,^2P_{3/2}$ state. The laser pulses were polarized (a) parallel and (b) perpendicular to each other. [From Hellmuth *et al.* (1981).]

(Haroche, 1976). In addition, $\beta_4(t)$ has different signs for both cases, but goes to zero after half a period of the hyperfine frequency. A β_4 of zero means, that the electronic part of the wave function is isotropic. The evolution, therefore, shows nicely how the spatial anisotropy in the intermediate state oscillates.

First experiments on the angular distribution of photoelectrons in two-photon ionization of sodium via the $3\,^2P_{3/2}$ intermediate state have been performed by Duncanson et al. (1976), Strand et al. (1978), and Hansen et al. (1980). In these experiments the maximum time delay chosen between excitation and ionization was too short to observe the periodic variation shown in Fig. 11. It has been shown, however, that the duration of the laser pulses used was not negligible compared to the hyperfine period, resulting in a reduction of the anisotropy of the electron angular distribution. This change of the angular distribution owing to the laser pulse duration and the quantum beats is also responsible for the differences in the angular distributions shown in Figs. 6 and 7.

In a similar way, the laser pulse duration may also affect the photoelectron spin polarization and the ratio of ion count rates for linearly and circularly polarized light. Such experiments have been performed by Granneman et al. (1977) in resonant two-photon ionization of cesium via either of the $7\,^2P$ states. For a further discussion, see Nienhuis et al. (1978).

6. *Ac Stark Effect*

It is well known that the ac Stark effect influences multiphoton ionization rates (Georges and Lambropoulos, 1977; Mainfray and Manus, 1980; and references therein). The experiments by Wheatley et al. (1978) and Agostini et al. (1978) on three-photon ionization of potassium and sodium show that the ratio of the ion signals for linear and circular laser polarization depends sensitively on the saturation of the bound–bound transition.

It is quite clear that the line shift induced by the ac Stark effect must also influence the angular distribution of photoelectrons. A suitable system to observe this effect is the three-photon ionization via the $3s \rightarrow 4d$ two-photon resonance in sodium. This has been studied theoretically by Dixit and Lambropoulos (1981). With the dye-laser frequency tuned to the $3\,^2S_{1/2}$, $F = 2 \rightarrow 4\,^2D_{5/2}$ transition, and for a low laser intensity, the photoelectron angular distribution exhibits two small side lobes similar to the ones shown in Fig. 7. As the laser intensity increases, the angular distribution changes, according to the calculations, first to a single-side and then back to a double-side lobe structure. The interpretation is, that with increasing intensity the $^2D_{5/2}$ state is Stark-shifted out of resonance and the $^2D_{3/2}$ state into resonance. At an even higher intensity both states are out of resonance. Of course when using pulsed lasers to observe intensity dependent effects, the change of the angular distribution observed in an experiment may be less pronounced. Dixit and Lambropoulos (1983) have also shown how the laser bandwidth changes this intensity-dependent effect.

In a recent experiment by Ohnesorge et al. (1983) two-photon resonant, three-photon ionization of sodium atoms of a thermal beam has been observed. A problem was that for intensities low enough to avoid Stark shifts, the ionization signal was very poor. Since the intensity effect is caused by the bound–bound 2-photon resonance, it is not sensitive to the frequency of the ionizing laser. Therefore the atoms were irradiated by the time-delayed second harmonic pulse of a Nd:YAG laser, also used to pump the dye laser inducing the 3s → 4d transition. Figure 12a shows an electron angular distribution for which the dye laser was tuned into resonance with the $3\,^2S_{1/2}$, $F = 2 \rightarrow 4\,^2D_{3/2}$ two photon resonance. This single-side-lobed shape is the same as that being observed in resonant three-photon ionization of sodium via the $3\,^2P_{1/2}$ state (Leuchs et al., 1979b; see also Fig. 6). The reason is that even though the $n\,^2D_{3/2}$ and $n\,^2D_{5/2}$ states had not been resolved spectrally in the resonant three-photon ionization only the $n\,^2D_{3/2}$ state can be excited in an electric dipole transition when starting from the $3\,^2P_{1/2}$ state.

The angular distributions shown in Figs. 12b–12d have been obtained under the same experimental conditions as the one in Fig. 12a, only the laser intensity has been changed among 1, 3, 6, and 10 MW cm^{-2}. In Fig. 12d two side lobes are clearly visible. Again the interpretation is that owing to the laser intensity the $3\,^2S_{1/2}$, $F = 2 \rightarrow 4\,^2D_{3/2}$ transition is shifted out of resonance. At the high dye-laser intensity used for Fig. 12d, there was enough ionization signal produced by the dye-laser pulse alone. Therefore the second harmonic pulse of the Nd:YAG laser was not used in this case.

This result demonstrates how sensitive photoelectron angular distributions may be to the laser intensity. It also opens an interesting possibility for studying the change of the atomic structure when the atom is exposed to high-intensity laser fields.

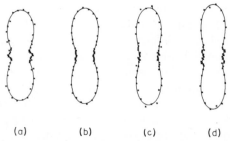

(a)   (b)   (c)   (d)

Fig. 12. The influence of the ac Stark effect on three-photon ionization of sodium via the $3\,^2S_{1/2}, F = 2 \rightarrow 4\,^2D_{3/2}$ two-photon resonance. The laser intensities are 1, 3, 6, and 10 MW cm^{-2} for (a)–(d), respectively. [From Ohnesorge et al. (1983).]

7. Off-Resonant Multiphoton Ionization

The term off-resonant shall be used for multiphoton ionization processes, where neither a bound state nor an autoionizing resonance is involved. The calculation of absolute cross sections for such higher-order off-resonant processes is extremely difficult, not only because of the infinite summations involved, but also because the atomic spectra are not known well enough and experimental data on high angular momentum states are often not available (Lambropoulos, 1976, 1980). Photoelectron angular distributions in off-resonant multiphoton ionization have been measured for the first time by Fabre *et al.* (1981) in Xe using frequency-doubled Nd:YAG-laser radiation.

A measurement of the angular distribution of photoelectrons in five-photon ionization of sodium atoms of a thermal beam has been performed by Leuchs and Smith (1982) using linearly polarized Nd:YAG-laser pulses at 1.064 μm of 12-ns duration and of 10^{10}-W cm^{-2} intensity. The smallest detuning between an atomic level and a multiple of the photon energy is 500 cm^{-1}, and at this intensity level, light shifts are much smaller than the detuning. The ionization process can therefore be considered off resonant. The resulting angular distribution is in agreement with the general shape suggested by Eq. (2.2). From a simple theoretical argument, one would expect in this case a vanishing cross section at $\Theta = 90°$ with respect to the polarization direction (Gontier *et al.*, 1975; Lambropoulos, 1976). However, this was not found experimentally. More detailed calculations by Dixit (1983) also were in considerable disagreement with the experiment.

In the experiment by Fabre *et al.* (1981), xenon atoms were ionized using linearly polarized frequency-doubled Nd:YAG-laser pulses. The absorption of six photons is enough to reach the $^2P_{3/2}$ continuum. Additional electron peaks in the spectrum of the electron kinetic energy are owing to the absorption of additional photons (Agostini *et al.*, 1981), a process referred to as continuum–continuum transition. The measurements show that the photoelectron angular distributions are changed distinctly by a continuum–continuum transitions. In this case, no theoretical results are yet available. Continuum–continuum transitions were also involved in photoelectron angular distribution measurements in five- and six-photon ionization of Xe via a three-photon resonance (Kruit, 1982).

III. CONCLUSION

Angular distributions of photoelectrons in resonant and off-resonant multiphoton ionization seems to be the catchword for a fast-growing field. Of course the question has to be asked, what new information about the physics of the atom and the ionization process can be obtained. The discussion in the preceeding sections has shown how sensitively photoelectron angular distributions depend on the bound- and continuum-state wave functions and thus on the one hand, on the atomic structure and on the other hand, on high-intensity effects in the atom field interaction. The higher multipoles appearing in the angular distributions allow the measurement of a larger number of independent parameters and should provide an improved basis for comparisons between theory and experiment. This should be especially true for the off-resonant multiphoton ionization discussed in Section II.B.7. There, an infinite number of atomic states contributes including those with an angular momentum quantum number as high as the number of photons necessary to ionize the atom. For a calculation of the cross section, a detailed and complete knowledge of the atomic spectrum is therefore indispensible.

Another aspect is that resonant multiphoton ionization may also yield information about photoionization cross sections out of excited atomic states. To date, only a few laboratory experiments on photoionization out of excited atomic states have been performed. Reliable cross sections for photoionization and recombination, however, are very important for the interpretation of astrophysical data. A severe problem in some experiments is the lack of knowledge about the degree of alignment or polarization of the atom in the excited state. It can be solved by either optically pumping the atom into only one magnetic sublevel (Smith et al., 1980) or by measuring the photoelectron angular distribution in addition to the absolute cross section.

Recently, several theoretical predictions on how high laser intensity may modify absorption and electron kinetic energy spectra in transitions to autoionizing states have been published (Lambropoulos, 1980; Lambropoulos and Zoller, 1981; Rzazewski and Eberly, 1981; Agarwal et al., 1982; Kim and Lambropoulos, 1982). Even though only preliminary experimental data exist to date (Feldmann and Welge, 1982), it can be expected that here also photoelectron angular distribution measurements will yield valuable information.

ACKNOWLEDGMENT

The financial support of the Deutsche Forschungsgemeinschaft is gratefully acknowledged. One of us (G. L.) would like to thank the Deutsche Forschungsgemeinschaft also for a Heisenberg Fellowship.

REFERENCES

Abragam, A., and Pound, R. V. (1953). *Phys. Rev.* **92**, 943.
Agarwal, G. S., Haan, S. L., Burnett, K., and Cooper, J. (1982). *Phys. Rev. Lett.* **48**, 1164.
Agostini, P., Georges, A. T., Wheatley, S. E., Lambropoulos, P., and Levenson, M. D. (1978). *J. Phys. B* **11**, 1733.

Agostini, P., Clement, M., Fabre F., and Petite, G. (1981). *J. Phys. B* **14**, L 491.
Alexandrov, E. B. (1964). *Sov. Phys. Opt. Spectros.* **16**, 522.
Amusia, M. Ya., Cherepkov, N. A., and Chernysheva, L. V. (1972). *Phys. Lett. A* **40A**, 15.
Arlanbekov, T. U., Grinschuk, V. A., Delone, G. A., and Petrosjan, L. B. (1975). *Sov. Phys. Lebedev Inst. Rep. (Engl. Transl.)* **10**, 33.
Arnous, E., Klarsfeld, S., and Wane, S. (1973). *Phys. Rev. A* **7**, 1559.
Berkowitz, J., and Ehrhardt, H. (1966). *Phys. Lett.* **21**, 531.
Bothe, W. (1924). *Z. Phys.* **26**, 59.
Cooke, W. E., and Gallagher, T. F. (1978). *Phys. Rev. Lett.* **41**, 1648.
Cooper, J., and Zare, R. N. (1969). *In* "Atomic Collision Processes, Lectures in Theoretical Physics XI-C" (S. Geltman, K. T. Mahanthappa, and W. E. Brittin, eds.), pp. 317–337. Gordon and Breach, New York.
Dixit, S. N. (1981). *J. Phys. B* **14**, L 683.
Dixit, S. N. (1983). *J. Phys. B* **16**, 1205.
Dixit, S. N., and Lambropoulos, P. (1981). *Phys. Rev. Lett.* **46**, 1278.
Dixit, S. N., and Lambropoulos, P. (1983). *Phys. Rev. A* **27**, 861.
Dodd, J. N., Kaul, R. D., and Warrington, D. (1964). *Proc. Phys. Soc., London* **84**, 176.
Ducas, T. W., Littman, M. G., and Zimmerman, M. L. (1975). *Phys. Rev. Lett.* **35**, 1752.
Duncanson, J. A., Strand, M. P., Lindgard, A., and Berry, R. S. (1976). *Phys. Rev. Lett.* **37**, 987.
Duong, H. T., Pinard, J., and Vialle, J.-L. (1978). *J. Phys. B* **11**, 767.
Edelstein, S., Lambropoulos, M., Duncanson, J. A., and Berry, R. S. (1974). *Phys. Rev. A* **9**, 2459.
Fabre, F., Agostini, P., Petite, G., and Clement, M. (1981). *J. Phys. B* **14**, L 677.
Feldmann, D., and Welge, K. H. (1982). *J. Phys. B* **15**, 1651.
Garton, W. R. S. (1978). *In* "Etats Atomiques et Moléculaires Couplés a un Continuum. Atomes et Molécules Hautement Excités" (S. Feneuille and J. C. Lehman, eds.). Edition du CNRS Nr. 273, Paris.
Georges, A. T., and Lambropoulos, P. (1977). *Phys. Rev. A* **15**, 727.
Georges, A. T., and Lambropoulos, P. (1978). *Phys. Rev. A* **18**, 1072.
Gontier, Y., Rahman, N. K., and Trahin, M. (1975). *J. Phys. B* **8**, L 179.
Granneman, E. H. A., Klewer, M., Nienhuis, G., and Van der Wiel, M. J. (1977). *J. Phys. B* **10**, 1625.
Hansen, J. C., Duncanson, J. A., Jr., Ring-Ling Chien, and Berry R. S. (1980). *Phys. Rev. A* **21**, 222.
Haroche, S. (1976). *In* "High Resolution Laser Spectroscopy" (K. Shimoda, ed.). Springer-Verlag, Berlin, Heidelberg, and New York.
Hellmuth, T., Leuchs, G., Smith, S. J., and Walther, H. (1981). *In* "Lasers and Applications," (W. O. N. Guimaraes, C. T., Lin, and A. Mooradian, eds.), Springer Ser. Opt. Sci., Vol. 26, pp. 194–203. Springer-Verlag, Berlin and Heidelberg.
Jacobs, V. L. (1972). *J. Phys. B* **5**, 2257.
Jacobs, V. L. (1973). *J. Phys. B* **6**, 1461.
Kaminski, H., Kessler, J., and Kollath, K. J. (1980). *Phys. Rev. Lett.* **45**, 1161.
Kim, Y. S., and Lambropoulos, P. (1982). *Phys. Rev. Lett.* **49**, 1698.
Klar, H. (1980). *J. Phys. B* **13**, 3117.
Klar, H., and Kleinpoppen, H. (1982). *J. Phys. B* **15**, 933.
Klarsfeld, S., and Maquet, A. (1972). *Phys. Rev. Lett.* **29**, 79.
Knight, P. L. (1979). *Opt. Commun.* **31**, 148.
Kruit, P. (1982). Ph.D. Thesis, FOM Institute for Atomic and Molecular Physics, Univ. of Amsterdam, Netherlands.
Kollath, K. J. (1980). *J. Phys. B* **13**, 2901.
Lange, W., and Mlynek, J. (1978). *Phys. Rev. Lett.* **40**, 1373.
Lambropoulos, M. M., and Berry, R. S. (1973). *Phys. Rev. A* **8**, 855.
Lambropoulos, P. (1972a). *Phys. Rev. Lett.* **28**, 585.
Lambropoulos, P. (1972b). *Phys. Rev. Lett.* **29**, 453.

Lambropoulos, P. (1973). *J. Phys. B* **6B**, L 319.

Lambropoulos, P. (1976). *Adv. At. Mol. Phys.* **12**, 87.

Lambropoulos, P. (1980). *Appl. Opt.* **19**, 3926.

Lambropoulos, P., and Teague, M. R. (1976). *J. Phys. B* **9**, 587.

Lambropoulos, P., and Zoller, P. (1981). *Phys. Rev. A* **24**, 379.

Leuchs, G. (1983). *In* "Laser Physics" (J. D. Harvey and D. F. Walls, ed.), pp. 174–194, Lecture Notes in Physics Vol 182, Springer-Verlag Berlin, Heidelberg, New York, and Tokyo.

Leuchs, G., and Smith, S. J. (1982). *J. Phys. B* **15**, 1051.

Leuchs, G., and Smith, S. J. (1983). *Proc. School Laser Appl. At. Mol. Nucl. Phys.*, *2nd, Vil'nius* (V.S. Letokhov, ed.), pp. 12–23. Institute of Spectroscopy, USSR Academy of Sciences, Moscow.

Leuchs, G., and Walther, H. (1977). *In* "Laser Spectroscopy III", (J. L. Hall and J. L. Carlsten, eds.), Springer Ser. Opt. Sci. Vol. 7, pp. 299–305. Springer-Verlag, Berlin, Heidelberg and New York.

Leuchs, G., Smith, S. J., Khawaja, E. E., and Walther, H. (1979a). *Opt. Commun.* **31**, 313.

Leuchs, G., Smith, S. J., and Walther, H. (1979b). *In* "Laser Spectroscopy IV" (H. Walther and K. W. Rothe, eds.), Springer Ser. Opt. Sci. Vol. 21, pp. 255–263. Springer-Verlag, Berlin, Heidelberg and New York.

Leuchs, G., Smith, S. J., Hellmuth, T., and Walther, H. (1980). Paper presented at the Lasers Conference, New Orleans, Louisiana, December, 15-19, 1980.

Leuchs, G., Reif, J., and Walther, H. (1982). *Appl. Phys. B* **28**, 87.

Leuchs, G., Matthias, E., Elliott, D. S., Smith, S. J., and Zoller, P. (1983). *In* "Laser Spectroscopy VI" (H. P. Weber and W. Lüthy, eds.) Springer Ser. Opt. Sci., in press. Springer-Verlag, Berlin, Heidelberg, and New York.

Mainfray, G., and Manus, C. (1980). *Appl. Opt.* **19**, 3934.

Matthias, E., Zoller, P., Elliott, D. S., Piltch, N. D., Smith, S. J., and Leuchs. G. (1983). *Phys. Rev. Lett.* **50**, 1914.

Mizuno, H. (1973). *J. Phys. B* **6**, 314.

Nienhuis, G., Granneman, E. H. A., and Van der Wiel, M. J. (1978). *J. Phys. B* **11**, 1203.

Ohnesorge, W., Diedrich, F., Elliott, D. S., Leuchs, G., and Walther, H. (1983). (To be published.)

Olsen, T., Lambropoulos, P., Wheatley, S. E., and Rountree, S. P. (1978). *J. Phys. B* **11**, 4167.

Rzazewski, K., and Eberly, J. H. (1981). *Phys. Rev. Lett.* **47**, 408.

Samson, J. A. R. (1982). *In* "Korpuskeln und Strahlung in Materie I" (S. Flügge and W. Mehlhorn, eds.), Handbuch der Physik, Vol. XXXI, Springer-Verlag, Berlin, Heidelberg, and New York.

Sandner, W., Kachru, R., Safinya, K. A., Gounand, F., Cooke, W. E., and Gallagher, T. F. (1983). *Phys. Rev. A* **27**, 1717.

Siegel, A., Ganz, J., Bussert, W., Hotop, H., Lewandowski, B., Ruf, M.-W., and Waibel, H. (1982). *Int. Conf. At. Phys., 8th, Göteborg.*

Siegel, A., Ganz, J., Bussert, W., and Hotop, H. (1983). (To be published.)

Smith, A. V., Goldsmith, J. E. M., Nitz, D. E., and Smith, S. J. (1980). *Phys. Rev. A* **22**, 577.

Sommerfeld, A., and Schur, G. (1930). *Ann. Phys. (Leipzig)* **4**, 409.

Strand, M. P., Hansen, J. C., Ring-Ling Chien, and Berry, R. S. (1978). *Chem. Phys. Lett.* **59**, 205.

Teague, M. R., and Lambropoulos, P. (1976). *J. Phys. B Phys.* **9**, 1251.

Tully, J. C., Berry, R. S., and Dalton, B. J. (1968). *Phys. Rev.* **176**, 95.

Wentzel, G. (1927). *Z. Phys.* **41**, 828.

Wheatley, S. E., Agostini, P., Dixit, S. N., and Levenson, M. D. (1978). *Physica Scripta* **18**, 177.

Yang, C. N. (1948). *Phys. Rev.* **74**, 764.

Zernik, W. (1964). *Phys. Rev. A* **135**, 51.

Zygan-Maus, R., and Wolter, H. H. (1978). *Phys. Lett. A* **64**, 351.

6

Above-Threshold Ionization: Multiphoton Ionization Involving Continuum–Continuum Transitions

P. AGOSTINI, F. FABRE, and G. PETITE

Service de Physique des Atomes et des Surfaces
Centre d'Etudes Nucléaires de Saclay
Gif-sur-Yvette, France

I. INTRODUCTION

Experimental studies on multiphoton ionization were initiated in the midsixties, only a few years after the Q-switched laser became available (Voronov and Delone, 1965; Agostini *et al.*, 1968). Since this pioneering work, most of the experimental effort has been focused on the behavior of the total cross section as a function of a variety of laser or atomic parameters (Lambropoulos, 1976; Morellec *et al.*, 1982). It is relatively recently that electron energy and angular distributions have been measured. Recent results in angular distributions are reported in Chapter 5 in this book by Leuchs and Walther and will not be commented on here. This chapter deals mainly with the energy spectrum of electrons ejected in multiphoton ionization of atoms, i.e., under high laser intensities.

According to the basic law of the multiphoton photoeffect, the electron energy is given by

$$E = N\hbar\omega - E_0 \tag{1.1}$$

where N is the minimum number of photons of energy $\hbar\omega$ required to ionize an atom with an ionization potential E_0. In a few experimental investigations (Martin and Mandel, 1976; Hollis, 1978; Boreham and Hora, 1979) it was reported that the photoelectrons had kinetic energies much larger than allowed by (1.1). Two groups of interpretations of this fact can be given: the fast electrons are produced either in the ionization process itself (one step) or in some secondary process following the ionization (two steps). The simplest among the one-step processes is an $(N + 1)$-ionization: a direct coupling between the ground state of the atom and the continuum by absorption of one photon more than strictly necessary for the ionization. This process, generalized to $(N + S)$-ionization (Gontier and Trahin, 1980) and named above-threshold ionization (ATI), is illustrated in Fig. 1. The energy spectrum it implies is, clearly, a series of lines separated by the photon energy $\hbar\omega$. The ejected electron energy is given by

$$E = (N + S)\hbar\omega - E_0 \tag{1.2}$$

Among the two-step processes that can "heat" the free electrons once they are produced by multiphoton ionization are the gradient force (or ponderomotive force) (Kibble, 1966; Boreham and Hora, 1979) and the inverse bremsstrahlung (Gavrila and Van der Wiel, 1978). The first one is a direct consequence of the space–time-dependent Lorentz force applied to the electron by the space–time-dependent electric field in a strongly focused laser beam. This force has been often invoked to explain the experimental results just mentioned. The second one is the absorption of one (or several)

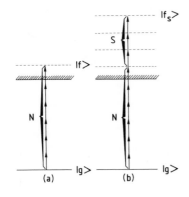

Fig. 1. Schematic diagram of ATI showing (a) the number N of photons just necessary to ionize and (b) the number S of photons absorbed above the ionization threshold. The ground and final states are $|g\rangle, |f\rangle$, and $|f_s\rangle$, respectively.

photons during the collision of the free electron with some target (ion or atom) (Weingartshofer *et al.*, 1979). This process also implies an energy spectrum, which is formed by a series of lines, identical to the one of ATI. A careful experimental test is therefore needed to distinguish between the two.

It was the aim of the work reported here to investigate experimentally the energy spectrum and angular distributions of electrons produced in multiphoton ionization. Since it was found that, under typical experimental conditions, ATI was the dominant process, ATI will be the main topic of this paper. From the theoretical point of view, the ATI process is obviously connected to multiphoton ionization but, more generally, to the problem of the decay of a discrete state coupled to multiple continua. It is also connected to the standard problem, in perturbation theory, of calculating a transition at some order when a lower order is allowed (Shiff, 1955). From the experimental point of view, it is related to multiphoton ionization techniques but also to photoemission experiments.

In Section II the theory of ATI is reviewed, stressing a few essential questions. Section III is devoted to experimental problems: the description of the techniques and discussion of the best experimental conditions for ATI; experimental study of ATI in xenon atoms at 1.06 and 0.53 μm. These results are discussed in Section IV within the framework defined in Section II.

II. ABOVE-THRESHOLD IONIZATION THEORY

The physical situation that has to be described is an atom in the ground state that is submitted to an electromagnetic field strong enough to induce, at least, multiphoton ionization. We are interested in the final states of the system where the ion is in its ground state and the free electron has a kinetic energy given by (1.2). Figure 2 shows, schematically, such transitions for xenon since the experimental results we are presenting have been obtained with this gas. For this specific case, the ground states of the ion ($5s^2 5p^5 \, ^2P_{1/2}$ and $5s^2 5p^5 \, ^2P_{3/2}$) have a fine structure splitting of 1.3 eV. It is reasonable to assume that excited states of the ion cannot be reached since, for xenon, the first excited state of the ion ($5s5p^6 \, ^2S$) has an energy of about 10 eV above the ion ground state. The experimental spectrum in the insert, showing negligible components around this energy, supports this assumption. In other words, we assume that only one outer electron of the atom is involved in the ATI process. The use of a hydrogen model is therefore expected to give reasonable predictions (at least qualitatively). Historically, the first theoretical works on ATI were dealing with a perturbative calculation of the cross section of hydrogen. More recently, a nonperturbative treatment

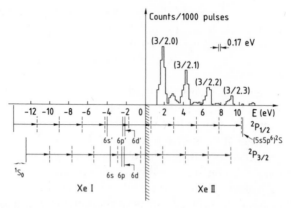

Fig. 2. Schematic energy level diagram of xenon. The energy origin has been taken at zero kinetic energy for the free electrons. The insert shows the corresponding energy spectrum of ejected electrons.

(Crance and Aymar, 1980a) discussing the validity of perturbation theory was published. Here we will summarize these works in the opposite order for the sake of clarity, and try to answer the following questions:

(i) What is the role of continuum–continuum transitions taking place in ATI? Is the population of some state, in the continuum, reached by ATI, dependent on the populations of other continuum states? What is the order of magnitude of this "free–free" coupling and can it be derived from the experiment?

(ii) What is the order of magnitude of the $(N + S)$-ATI cross section compared to the one of a "normal" multiphoton ionization involving the same number of photons? In this respect do the intermediate resonant continua play a role similar to that played in multiphoton ionization by discrete intermediate resonant states?

A. Nonperturbative Treatment of ATI

The problem of the evolution of a discrete state coupled to a set of continua has been abundantly treated in the past. This situation is encountered in many different physical systems like the decay of elementary particles, nonradiative decay of small molecules, photochemical reactions. Various techniques have been used to solve the problem. One of the most attractive, the Green's function method, has been used by Nitzan *et al.* (1973). Using a very general scheme, they were able to obtain the probabilities of finding, at time t in the continuum, a system initially in the discrete state, and the branching ratio for the populations of the different continua. Here we will follow a more recent treatment (Crance and Aymar, 1980a) more specifically describing ATI. Let us note here that a complete analogy between these two calculations can be easily proved and that their results and approximations are equivalent.

Using the Crance and Aymar notation, the state of the system at time t can be expanded as

$$|\Psi(t)\rangle = a_0(t)|0\rangle + \sum_{K,\alpha} \int_{k\hbar\omega-\varepsilon-E_0}^{k\hbar\omega+\varepsilon-E_0} dE_k^\alpha \, a_{E_k^\alpha} \exp(-ik\omega t)|E_k^\alpha\rangle \qquad (2.1)$$

where $|0\rangle$ is the atomic ground state, $|E_k^\alpha\rangle$ is the atomic continuum state with energy E_k, α labels the continua with same energy but different quantum numbers, $K = N + S$, and ω is the light frequency.

By using an effective Hamiltonian H_{eff} acting on the restricted basis $|0\rangle$, $|E_k^\alpha\rangle$ the Schrödinger equation can be written as

$$i\dot{a}_0 = \sum_{k,\alpha} \int_{k\hbar\omega-\varepsilon-E_0}^{k\hbar\omega+\varepsilon-E_0} dE_k^\alpha \, K_k^\alpha(E_k^\alpha) a_{E_k^\alpha} \qquad (2.2)$$

$$i\dot{a}_{E_k^\alpha} = \delta_{E_k^\alpha} a_{E_k^\alpha} + K_k^\alpha(E_k^\alpha) a_0$$
$$+ \sum_{k',\alpha} \int_{k'\hbar\omega-\varepsilon-E_0}^{k'\hbar\omega+\varepsilon-E_0} dE_{k'}^{\alpha'} \, L_{kk'}^{\alpha\alpha'}(E_k^\alpha, E_{k'}^{\alpha'}) a_{E_{k'}^{\alpha'}} \qquad (2.3)$$

where K_k^α and $L_{kk'}^{\alpha\alpha'}$ are the transition matrix elements between the ground state and the continuum and between continuum states, respectively.

By making two basic approximations, which are discussed hereafter, the following equations governing the populations of the ground state and the continua can be obtained:

$$\dot{n}_0 = -\sum_{k,\alpha} 2\pi |N_k^\alpha|^2 n_0 \qquad (2.4)$$

$$\dot{\mathcal{N}}_k^\alpha = 2\pi |N_k^\alpha|^2 n_0 \qquad (2.5)$$

where

$$N_\alpha^k = \sum_{k'\alpha'} [(1 + i\pi\mathcal{L})^{-1}]_{kk'}^{\alpha\alpha'} K_{k'}^{\alpha'} \qquad (2.6)$$

$\mathcal{L}$ is the matrix defined by the $L_{kk'}^{\alpha\alpha'}$. $|N_\alpha^k|^2$ must be interpreted as the probability per unit time for the ground state to ionize to the continuum $|E_k^\alpha\rangle$. Let us stress here the consequences of (2.4) and (2.5): (i) the ionization towards several continua is described by rate equations coupling each continuum population to the ground-state population and not to other continua populations, although coupling between continua do exist through the matrix $\mathcal{L}$. This result is identical to the one derived by Nitzan et al. (1973); (ii) the intensity dependence of the rates $|N_\alpha^k|^2$ is included in the intensity dependences of $\mathcal{L}$ and $K_{kk'}^{\alpha\alpha'}$. Therefore, in general, it will be different from the prediction of the lowest-order perturbation theory since

$$|K_k^{\alpha'}|^2 \sim I^{N+S}$$

and

$$(|L_{kk'}^{\alpha\alpha'}|^{-1})^2 \sim I^{-|k-k'|}$$

where I is the intensity of the electromagnetic field.

For small values of $L_{kk'}^{\alpha\alpha'}$ (which are encountered for real atoms), it is easily shown that the rates $|N_\alpha^k|^2$ behave like I^{N+S}, while for large values of $L_{kk'}^{\alpha\alpha'}$ it would behave like $I^{N'}$ with $N' < N + S$; (iii) the rates are time dependent through their intensity dependence. For some pulse shapes the system (2.4), (2.5) can be analytically solved but, in general, it will have to be numerically integrated. This is done in Section III to fit experimental data. It will be shown then, that the apparent "slope" ($d \log \mathcal{N}_k^\alpha / d \log I$) can be lowered by the saturation, i.e., the depletion of the ground state, even if the $L_{kk'}^{\alpha\alpha'}$ are vanishingly small.

Let us discuss briefly the approximations on which the system of Eqs. (2.4) and (2.5) is based.

The first one is that ε [in (2.1)] has to be much larger than the energy spread of each peak in the energy spectrum. This implies, as mentioned by Crance and Aymar (1980a), that the characteristic evolution time of the continuum is very small and that the balance between states in different continua is immediately reached.

The second is that K_k^α and $L_{kk'}^{\alpha\alpha'}$ are energy independent. It implies that the density of states over ε can be considered constant. Let us note that this approximation is the basic one in the Nitzan et al. (1973) treatment.

We can summarize this section by answering some of the questions raised in the introductory remarks:

The continuum–continuum coupling (or free–free transitions) modify the rate governing the population of continua $\mathcal{N}_k^\alpha$, but $\mathcal{N}_k^\alpha$ does not depend on the $\mathcal{N}_k^\alpha$.

This coupling is usually very small, at least for not too high intensities. It could be modifying the intensity dependence of the rates. However it cannot be derived directly from slope measurements but only from absolute measurements of the probabilities.

To conclude this part, another key result of the Crance and Aymar calculation should be pointed out, namely that the maximum branching ratio for the populations of the various continua is reached for the saturation intensity I_s, which can be defined by

$$\sum_{s=0}^{\infty} \sigma_{N+S} I^{N+S} \tau \simeq 1 \tag{2.7}$$

where σ_{N+S} is the lowest-order N-photon ionization cross section and τ is the interaction time.

From the above discussion it can be concluded that in a rather large range of intensities perturbation theory will give correct predictions for the branching ratios and the slopes. However, if the pulse peak intensity I_M is larger than the saturation intensity I_S, the latter should be used in the calculation. Under these conditions the usual lowest-order approximation for $H_{\rm eff}$ might be questionable. This point is discussed in the next section.

B. Perturbative Calculations of ATI Probabilities

The effect of higher-order terms in perturbation theory has been studied by Gontier *et al.* (1976) and more recently by Aymar and Crance (1981). The latter show that, in order to get consistent results, the ionization probabilities intensity expansion $P_N, P_{N+1}, \ldots, P_{N+S}$ should be calculated up to $(N + S)$-order. For example, the 2-photon ionization probability is

$$P_2 = \sigma_2^1 I^2 + (\sigma_2^2 - \lambda\sigma_2^1)I^3 \tag{2.8}$$

where σ_2^1 is the lowest-order 2-photon cross section, and σ_2^2 is the next order in the perturbation expansion given by

$$\sigma_2^2 = 2(a + b + c + d + e) \tag{2.9}$$

where a, b, c, d, and e are the amplitudes corresponding to the various possible quantum paths involving six interactions and the net absorption of two photons.

Numerical results for cesium and potassium showing this correcting terms are shown in Table I, reproduced from Aymar and Crance (1981). The corrections for intensities up to $I \sim 10^2$ GW cm^{-2} are small; they do not exceed 15% of the main term but cannot be neglected. The influence of these terms on the slope $[d(\log P_2)/d(\log I)]$ is not measurable.

Table I

Ionization Probabilities per Unit Time (s^{-1}) for 2-, 3-, and 4-Photon Ionization of Cs and K at 0.53 μm[a,b]

Cs	P_2	$0.104 \times 10^7 I^2 - 0.15 \times 10^3 I^3$
	P_3	$0.795 \times 10^2 I^3$
	P_4	$0.197 \times 10^{-2} I^4$
K	P_2	$0.178 \times 10^7 I^2 - 0.37 \times 10^3 I^3$
	P_3	$0.171 \times 10^3 I^3$
	P_4	$0.545 \times 10^{-2} I^4$

[a] Reprinted with permission from Aymar and Crance (1981). Copyright 1981 by The Institute of Physics.
[b] I is in GW cm^{-2}.

C. Numerical Calculations

Within the validity limits discussed above, the $(N + S)$ ionization rate can be calculated by Fermi's "golden rule" and, therefore is proportional to the square of an $(N + S)$-order matrix element, the computation of which is the hard matter of the problem, especially because of the divergences that appear in some of the infinite summations. The first computation was performed by Zernik and Klopfenstein (1965) for $N = 1$ and $S = 1$ in hydrogen. They used the implicit summation technique of Dalgarno and Lewis (1955) and, by a suitable integration in the complex plane, they were able to remove the divergence and to obtain numerical values for the 2-photon ATI cross section. One conclusion can be drawn at once from their results: the cross section is quite comparable to the nonresonant, under-threshold, 2-photon probability. In other words, nothing dramatic happens because of the "resonance" in the continuum. This is at variance with the usual discrete state resonances that give rise to infinite values and to a drastic increase of the cross section.

This "smooth" behavior is due to the following reason. When a resonance occurs in the continuum, the corresponding energy denominator appears in an integral, the contribution of which is a Cauchy principal value and therefore small. Physically this is connected to the fact that the lifetime of a continuum state is much shorter than the lifetime of a bound state since the electron (which has a kinetic energy of the order of 1 eV) very rapidly leaves the vicinity of the ion. The order of magnitude of this lifetime (10^{-13} s) is thus much shorter than for a bound state (10^{-8} s).

The Zernik calculation has, since then, been confirmed and extended by a number of authors (Klarsfeld, 1970; Karule, 1971, 1976; Klarsfeld and Maquet, 1979a,b; Gontier and Trahin, 1980). The reader is referred to these works for details on the various techniques: Sturmian expansion, Pade approximant, generalized implicit summation techniques. The highest-order cross section has been given by Gontier and Trahin (1980): $N = 6$, $S = 6$, $(N + S = 12)$ for the ground state of hydrogen. Table II shows their results. The quantities tabulated are the generalized ionization cross sections (Lambropoulos, 1976; Morellec et al., 1982) σ_{N+S}, an intensity independent quantity allowing us to write the $(N + S)$ ionization rate as

$$W^{(N+S)} = \sigma_{N+S} I^{N+S} \tag{2.10}$$

Several conclusions can be derived from this table. First, by comparing in two different columns the σ_{N+S} for different values of N and S but for identical $N + S$, we note that they have almost the same order of magnitude. Second, in each column we note that the ratio of two successive order σ_{N+S} has a constant value (approximately 10^{-14} W^{-1} cm^2).* This had al-

---

* This ratio has the dimension of the inverse of an intensity. Expressed in centimeter squared seconds, it is equal to 4×10^{-33} for 2.34-eV photons.

Table II

Calculated Values for the $(N + S)$-Photon Ionization Generalized Cross Section[a] for the Hydrogen Atom in the Ground State[b]

S	N			
	6 (5300 Å)	8 (6500 Å)	10 (9100 Å)	12 (10 820 Å)
0	1.39 (-69)	1.49 (-97)	4.51 (-123)	3.46 (-149)
1	2.84 (-83)	9.85 (-111)	7.78 (-136)	9.81 (-162)
2	2.92 (-97)	2.53 (-124)	5.35 (-149)	1.10 (-174)
3	2.80 (-111)	5.84 (-138)	2.61 (-162)	1.08 (-187)
4	2.66 (-125)	1.35 (-151)	1.89 (-175)	9.87 (-201)
5	2.32 (-139)	2.75 (-165)	1.04 (-188)	8.91 (-214)

[a] Values calculated in $s^{-1} W^{-(N+S)} cm^{2(N+S)}$.
[b] The numbers in parentheses indicate powers of 10.

ready been noted in the case of "normal" multiphoton ionization processes (Morellec *et al.*, 1982) and the numerical values of these ratios are the same (when the same units are used for σ_{N+S}). This supports the following conclusions concerning the second set of question raised in the introducing remarks: the cross sections of $(N + S)$-ATI are quite "normal." The intermediate resonant continua do not increase the transition probability like a discrete resonant state. Let us point out here that, if a resonance with a discrete state occurs in an ATI, the corresponding cross-section will be increased like a "normal" multiphoton ionization cross section. For alkali atoms the numerical calculations rely on a model potential (Crance and Aymar, 1980b) and have been carried out by Aymar and Crance (1981), including the higher-order corrections mentioned in the previous section.

III. ABOVE-THRESHOLD IONIZATION EXPERIMENTS

Experimental investigation of ATI began only recently, mainly because of the many difficulties met in such experiments and primarily because, as will be proved later, the use of electron counting methods is almost compulsory. This means that these experiments were made possible only by the use of high-power high-repetition-rate laser sources which have only recently been available.

In this section, we shall review the different electron spectroscopy methods and the different lasers used in ATI experiments. We then consider the question of the various experimental conditions (gas pressure, laser intensity, electron collection, etc.). Finally we present and discuss the experimental results.

A. Electron Spectroscopy

Any of the classical methods of electron spectroscopy can be used in ATI experiments. Most of them are well known and have been recently reviewed by Ballu (1980) and will be quickly surveyed here because they may not be familiar to most multiphoton physicists.

1. Nondispersive Methods

Nondispersive methods are those in which the position of the electron at the collector does not depend on the electron energy. Two of the methods that have been used are the retarding-potential method and the time-of-flight method.

In the retarding-potential method, the electrons are decelerated by an adjustable potential and only those having a high enough kinetic energy can reach the detector. The electron energy spectrum is thus obtained as the derivative of the electron signal dependence towards the retarding potential. Though retarding-potential methods with a careful electrode design can reach a high degree of accuracy, only their roughest version has been used in ATI experiments (Ballu, 1980).

Time-of-flight spectroscopy has been long used in multiphoton ionization to isolate a given ion signal from parasitic signals (impurities, surface ions, molecular ions, etc.). Here the (kinetic) energy selection is obtained by letting the electrons (or the ions) propagate in a field–face space. The energy distribution is thus translated into a time distribution of the electron (or ion) signal on the detector. In ion measurement experiments, the kinetic energy distribution reflected the mass differences between different ions accelerated under the same potential (Morellec et al., 1982). In an ATI experiment this energy distribution simply reflects the spectrum of the different final states, whether the electrons are accelerated or not. The most exciting feature of this method is that it is a multichannel method: a single laser shot yields informations on the complete electron spectrum. Its worst drawback is the decreasing selectivity for increasing electron energies.

2. Dispersive Methods

In these methods, electrons with different energies are spatially separated by a suitable electric or magnetic field configuration, and the energy selection is made by use of a pinhole, slit, or whatever device is suitable to the apparatus geometry. Many different electron spectrometer designs have been used (Ballu, 1980), but only the 180° spherical sector analyzer was used in ATI experiments. It consists of two metallic hemispheres between which the electron beam passes. Electron trajectories are curved between the two spheres and the energy dispersed. Moreover, if the beam enters the intersphere space through a pinhole, it is automatically refocused at a 180° deflection angle at a point whose position depends on the electron energy. 180° spherical sector analyzers are famous because of their very high selectivity, which reaches the meV range. The energy spectrum can be scanned either by scanning the voltage difference between the internal and external hemispheres or by keeping this difference constant while simultaneously scanning both voltages with respect to the interaction region potential taken as the electrical ground. This last operating mode has the advantage of keeping the energy resolution constant and is generally preferred.

Special electron-optics designs have to be used to image the interaction volume on the input pinhole with minimum chromatic aberration so that the apparatus transmission does not suffer large variations when the electron energy spectrum is scanned.

B. The Lasers

Given the intensity range at which ATI processes are expected to be observed (10^{13}–10^{14} W cm^{-2} for hydrogen atoms), there is no doubt as to why, thus far, only high-power solid-state lasers have been used. As a matter of fact, all ATI experiments but one have been performed using neodymium lasers operated in the nanosecond pulse length range. Typical characteristics of such lasers are an energy per pulse within a few tenths of a joule to 3 J, with pulse length ranging from 15–30 ns. After focusing, such lasers can deliver intensities up to 10^{13} W cm^{-2}.

One experiment (Kruit et al., 1981) used a Nd:YAG-pumped dye laser. Though intensities available with such lasers are not as high as those obtained with Nd lasers, their tunability allows the use of resonances on intermediate states to increase the ionization probability. Intensities between 10^{10} and 10^{11} W cm^{-2} were obtained with such lasers. Recent development will render possible the use of high-power excimer laser (uv) in this field.

C. Experimental Conditions

Detection of low-energy electrons brings about many problems which were not met in the case of ion detection. If one is not interested in angular distribution, it is possible to use an accelerating field, which should be much smaller than those used in ion detection. One should, however, be careful that this does not only result in a translation of the energy spectrum, but also in a broadening of individual peaks. This broadening finds its origin in the finite dimension of the interaction volume and is generally very small. A very interesting system has been described by Kruit and Read

Table III

Debye Lengths (in cm) for Different Electron Energies
and Different Gas Pressures (Atomic Densities)

	E (eV)		
N (cm^{-3})	0.1	1	10
10^{10}	1.9 10^{-3}	6.1 10^{-3}	1.9 10^{-2}
10^{11}	6.1 10^{-4}	1.9 10^{-3}	6.1 10^{-3}
10^{12}	1.9 10^{-3}	6.1 10^{-4}	1.9 10^{-3}

(1983) that allows collection on a 2π solid angle without loss of selectivity. This device, called a magnetic mirror, works on a principle identical to that of the magnetic bottle and has been used in connection with a time-of-flight spectrometer, reaching a resolution of 15 meV. When either no collection field or a magnetic field is used, one has to face the problem of the extraction of low-energy electrons from a plasma. This problem can be considered in the framework of ambipolar diffusion theory, though it is valid in the case of plasma at thermal equilibrium, which is not the case here. Total extraction of the electrons of a given energy is possible only if the corresponding Debye length λ_D is large compared to the plasma dimensions, in our case the interaction volume dimensions. Table III gives the value of λ_D (cm) for different electron energies and different gas pressures (plasma densities). Plasma dimensions and plasma densities are determined by the diameter of the interaction volume and the gas pressure (Morellec *et al.*, 1982). Interaction volumes have diameters of the order of about 10 μm. From Table III it is clear that extraction of electron with an energy of 0.1 eV will be very difficult without an accelerating field, even for very low pressures (10^{10} cm^{-3} correspond to a pressure of 3 10^{-7} Torr.). Even detection of 1-eV electrons requires working at a very low gas pressure, which in turn decreases the counting rate.

Another effect of increasing pressure is that electrons that have left the interaction volume are decelerated by the positive charge they have left behind. Using a very rough model (one electron with an initial kinetic energy of a few electron volts submitted to the Coulomb force of a charge equal to the sum of all the ions' charges), this effect can be estimated to a few tenths of an electron volt for a pressure of 5×10^{-6} Torr.

All this discussion emphasizes the necessity of working at very low pressures when no accelerating field is applied.

D. Experimental Results

Experiments on ATI are still very few. Moreover, all the experiments reported so far have used xenon as a gas. This can be easily understood because atomic hydrogen is not easy to handle experimentally. On the other hand, Xe is the easiest of the common rare gases to ionize and its ionization threshold (12.127 eV) is close to that of hydrogen.

Three types of experiments at three different wavelengths have been performed:

    (i) 1.06 μm, with Nd: YAG or Nd:glass lasers,
    (ii) 0.53 μm, with the same lasers, but frequency doubled,
    (iii) 0.44 μm, with a pulsed dye laser, with a 3-photon resonance on the 6S ^{2}P$_{3/2}$ state of Xe.

The first experimental demonstration of ATI was given by Agostini *et al.* (1979). Energy spectra reported in this paper are shown in Fig. 3. They were obtained using a Nd:glass laser and a retarding potential analyzer. Both the fundamental and the second harmonic frequencies were used in this experiment. At the 0.53-μm wavelength the spectrum exhibits two peaks separated by about 2.4 eV, in good agreement with the photon energy at this wavelength (2.34 eV). Note that the intensity reported in this experiment is about 2×10^{12} W cm^{-2} and the direction of detection is perpendicular to the laser polarization.

At 1.06 μm no peak structure is observed and the spectrum peaks at about 4.5 eV. The authors suggest that this may be due to gradient forces, but it must be pointed out that these results were obtained at a pressure of

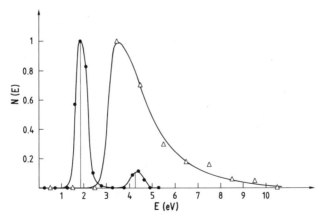

Fig. 3. Electron energy spectra by Agostini *et al.* (1979): ●, 0.53 μm; △, 1.06 μm.

10^{-4} Torr which is, given the discussion above, too high to allow a good collection of the outgoing electrons.

Even the 0.53-μm spectrum is not completely satisfactory. It should be remembered that the XeII ground state is a $5s^2 5p^5$ state, which can be either in a $P_{1/2}$ or a $P_{3/2}$ state (see Fig. 1). This leads to two different ionization thresholds, depending on the final state of the ion. As a consequence, two series of peaks should be observed, distant by the core fine structure splitting of 1.3 eV.

This experiment was therefore repeated with a more sophisticated equipment and better defined conditions. The results are published in Agostini *et al.* (1981) and Fabre *et al.* (1981, 1982a) and present the state of the art in the ATI field, and therefore will be discussed in full details.

These experiments first differed by the use of a spherical sector electron analyzer (see Section III.A.2). This apparatus was not used at its maximum selectivity to keep its transmission compatible with the counting rates required. However, when used at a 20-V analysis energy (energy of the average trajectory inside the analyzer), it gave a 0.1-eV theoretical resolution, which was sufficient to fully exhibit the expected structures.

Another difference is the use of a high repetition rate (10 Hz) Nd:YAG laser, and electron counting techniques. A detailed description of the equipment is given by Fabre *et al.* (1982). Figure 4 shows the experimental arrangement of the laser beam, the interaction volume, and the electron analyzer, which is a typical example of arrangements used in ATI experiments.

Two electron spectra obtained at 0.53 and 1.06 μm wavelengths are shown on Fig. 5. Counting rates in this experiment ranged between 10^{-3} and 1 count/laser shot. Accumulation lasted until about 100 electrons were counted at the maximum of the peaks. The spectrum at 0.53 μm shows two series of 3 peaks due to 6-, 7-, and 8-photon ionization of xenon, when

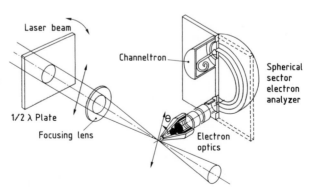

Fig. 4. Experimental apparatus of Fabre *et al.* (1981, 1982).

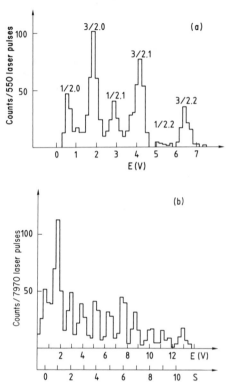

Fig. 5. Energy spectrum of the outgoing electrons for multiphoton ionization of xenon at (a) 0.53 μm, (b) 1.06 μm (Fabre *et al.*, 1982). Peaks of (a) are labeled (J/S) J, core total angular momentum; S, number of photons absorbed above the minimum.

only 6 photons are necessary to ionize xenon atoms at 0.53 μm. In each series, the peak spacing is in good agreement with the 2.34 eV photon energy. The two series are separated by the core fine structure splitting 1.3 eV. These results were obtained for a laser intensity of 10^{11} W cm^{-2} and a xenon pressure of 10^{-6} Torr.

The spectrum at 1.06 μm shows only one series of 11 evenly spaced peaks corresponding to 11–21-photon ionization of xenon, when 11 photons only are necessary to ionize xenon at that wavelength. Only one series is seen because the photon energy in this case is 1.17 eV, and the two series are separated by 1.3 eV only and cannot be seen with the resolution of this apparatus. This spectrum was obtained for an intensity of about 3×10^{12} W cm^{-2} and a pressure of 10^{-6} Torr.

The dependence of these spectra on the different parameters of the experiment was studied: pressure, laser intensity, laser polarization direction (the laser was linearly polarized).

The amplitude of the peaks of Fig. 5a was shown to depend linearly on the pressure between 5×10^{-7} and 10^{-5} Torr. This shows that the peak structure of the spectra is indeed due to ATI. Such a structure could have been explained by a stimulated inverse bremsstrahlung process (a laser photon is absorbed during the collision between an electron resulting from "pure" multiphoton ionization and a xenon atom or ion), but such a process would have led to a quadratic dependence of the signal toward pressure. A red shift and a broadening of each peak can be seen above 5×10^{-6} Torr and this broadening is such that the peak structure is washed out for pressures greater than 10^{-5} Torr because of space charge effects discussed in Section III.C.

The angular distribution of the outgoing electrons for the first two peaks of the series corresponding to the $^2P_{3/2}$ core configuration was studied by

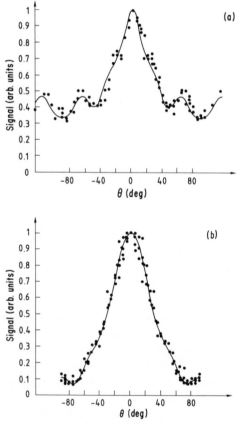

Fig. 6. Angular distribution of the outgoing electrons in $(6 + S)$ ionization of xenon at 0.53 μm (Fabre *et al.*, 1981). (a) $S = 0$, (b) $S = 1$.

Fabre *et al.* (1981) by rotating the laser polarization. Figure 6 shows these two angular distributions. We will only briefly comment this result since the question of angular distributions is discussed in Chapter 5 by Leuchs and Walther. Let us note that both distributions are peaked around the laser polarization and that the 6-photon distribution is broader than the 7-photon one. Another interesting result of Fabre *et al.* (1981) is that the spectrum obtained at 90° of the laser polarization looks like the one obtained in Agostini *et al.* (1979) under the same conditions. The series corresponding to the $^2P_{1/2}$ core configuration seems to have, in this case, a negligible contribution.

The intensity dependence of the peak amplitudes yields major informations about the fundamental questions raised in Section II of this chapter. This dependence is studied in Fabre *et al.* (1982) and Fig. 7 summarizes the results of their experiment concerning the first three peaks of the $^2P_{3/2}$ core configuration series. The three curves correspond to $S = 0, 1, 2$ (starting from the top). The points are experimental results and the solid lines are the results of a fit discussed in Section IV. When the slopes of the linear parts of these curves are measured, they are found somewhat lower than the expected theoretical slopes $(N + S)$ (Fabre *et al.*, 1982). However, we show in Section IV that this feature can be accounted for in the framework of the theory discussed in Section II. The observation of a significant contribution

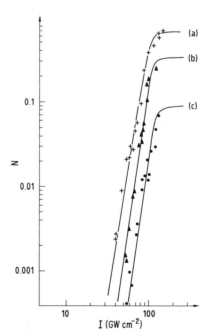

Fig. 7. Intensity dependence of the $(3/2, 0)$ (curve A), $(3/2, 1)$ (curve B), and $(3/2, 2)$ (curve C) peaks' amplitude (see Fig. 5a) in log–log scale. Points are experimental results. The solid line is the result of the fit discussed in Section IV.

P. AGOSTINI, F. FABRE, AND G. PETITE

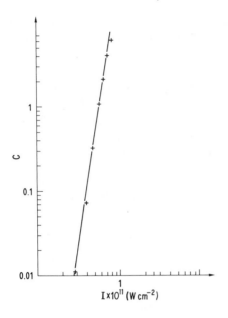

Fig. 8. Intensity dependence of the total electron signal (log–log scale).

of ATI to multiphoton ionization brings about an important question. If successive peaks of increasing order have intensity dependences steeper than that of the lowest-order peak, why has not this affected the intensity dependence of the total ion signal which, in earlier experiments, was measured with a good accuracy and found equal to the smallest number of photons necessary for ionization? Figure 8 shows the intensity dependence of the total electron signal, i.e., the sum of the contribution of all the significant peaks. The slope of the curve in Fig. 8, in log–log scale, is found to be 6.1 ± 0.3 for a 6-photon (minimum) process. This is in better agreement with the theoretical slope (value 6) than any other previous measurement reported in this case. This can be explained as follows: spectra like those of Figs. 5a and b have been obtained at high laser intensities where the effect of saturation cannot be neglected. Saturation mainly causes a decrease of the slope of curves like those of Figs. 7 and 8. On the contrary, ions slope measurements were made at laser intensities low enough so that saturation would not perturb the intensity dependence. As a matter of fact, spectra obtained in the linear part of the curves of Fig. 7 show that the main contribution to the electron production is due to the peak of minimum order. This is discussed at length in Fabre *et al.* (1983).

It is interesting to compare spectra from Figs. 5a and 5b. They differ in that the spectrum at 1.06 μm shows a much larger number of peaks than the one at 0.53 μm. The same conclusion can be drawn from results presented by Kruit *et al.* (1981). First, in view of the above discussion, let us note that results at 1.06 μm were obtained in the region of saturation of the ionization process. The main reason that spectra at 1.06 μm shows many more peaks than that at 0.53 μm is that the saturation intensity is higher in the 1.06 μm case. As shown in Section II, increasing the laser intensity enhances the amplitude of high-order peaks, but the saturation intensity sets a maximum value of the laser intensity beyond which this enhancement stops being effective. Thus the difference between the two spectra can be easily understood within the framework of the theory presented in Section II.

IV. DISCUSSION

Even though the case of xenon is far from being the most suitable for a comparison between theory and experiment, some valuable informations can be obtained from the experimental results of Section III.

First, they prove that ATI, and not stimulated inverse breemsstrahlung, is responsible for the energetic electrons detected in these experiments. This was expected because collisions cannot play a significant role at the very low pressures at which these experiments were performed. The pressure dependence of the different ATI peaks proves it unambiguously.

Second, despite the complexity of the xenon atom, some of the questions raised in Section II can be answered.

Most of the answers can be found in the intensity dependence of the peaks amplitude. The fact that this dependence, in the absence of saturation, is well represented by a straight line of slope $N + S$ seems to prove that (i) high-order contributions to the ionization rates are of minor importance if any of (ii) the same conclusions can be derived for the coupling between different continua. The effect of these two terms would have been to modify the slope of curves of Fig. 7.

Both the nature of the experimental results and the difficulties of multiphoton calculations in xenon prohibits a determination of absolute values of ATI cross sections. However, it is possible to obtain informations upon the relative magnitude of the ionization cross sections corresponding to the results of Fig. 7. The curves of Fig. 7, drawn in solid lines were obtained through numerical integration of the system of differential Eqs. (6)–(8). The following assumptions were made:

(i) In Eq. (10) $\mathscr{L}$ is supposed to be the 0 matrix, i.e., the continuum–continuum couplings are neglected.

(ii) Consequently, the $|N_k^\alpha|^2$ of Eqs. (6) and (7) are simply equal to $|K_k^\alpha|^2$, that is proportional to the ionization rates towards the three continua. Lowest-order approximations of these quantities were used.

(iii) A Gaussian pulse shape was assumed.

Thus the rates to the $S = 0, 1, 2$ continua were written under the form

$$S_0 I_M^6 \exp[-6(t/\tau)^2] \quad (S = 0)$$

$$S_1 I_M^7 \exp[-7(t/\tau)^2] \quad (S = 1)$$

$$S_2 I_M^8 \exp[-8(t/\tau)^2] \quad (S = 2)$$

where τ is the pulse duration, I_M is the pulse amplitude, and S_0, S_1, S_2 are quantities proportional to $\sigma_6, \sigma_7, \sigma_8$, respectively, the generalized ionization cross sections toward the three continua. For a given I_M, time integration was carried out for $-3\tau < t < 3\tau$ and I_M was then incremented. S_0, S_1, S_2 were then determined by fitting the curves obtained in this way to the experimental data.

The fit between experimental points and the result of such a calculation is satisfying and was found to be very sensitive to the relative magnitude of the ionization cross section, which we think are determined to be less than a factor of 2, notwithstanding the uncertainty on the experimental laser intensity.

Moreover, the above values of S_0, S_1, S_2 correspond to relative generalized cross sections of $\sigma_7/\sigma_6 = 1.5 \times 10^{-30}$ cm^2 s and $\sigma_8/\sigma_7 = 7.1 \times 10^{-31}$ cm^2 s. These figures are close to those obtained in the case of multiphoton ionization processes of successive orders, but involving only discrete intermediate states (Morellec et al., 1982). This supports the conclusion that ATI processes, in this intensity range, do not differ signicantly from "normal" multiphoton ionization processes on the question of the order of magnitude of the generalized cross sections. This contradicts the conclusions of Fabre et al. (1983) on this point, which were based on a comparison of processes of the same order, but in different gases.

As a conclusion to this discussion, it can be said that a model like the one discussed in Section II is adequate to account for all the experimental observations on ATI thus far. Concerning the continuum–continuum coupling, this experiment proves that it is small compared to all the other couplings of the problems, at least for xenon and laser intensities not higher than 10^{11} W cm^{-2}. For higher laser intensities or other atoms this coupling or electron correlation might have to be taken into account.

V. CONCLUSION

In this chapter we have reported theoretical and experimental results on above-threshold ionization of atoms under strong laser irradiation. The main conclusions of this work can be summarized as follows:

(i) Under typical experimental conditions, inverse bremsstrahlung can be ruled out and ATI is the process responsible for the production of fast electrons.

(ii) The observed spectra and their intensity dependence are well predicted by a model of a one-electron atom coupled to multiple continua, treated by perturbative methods. This implies two important consequences, at least for laser intensities less than of 10^{11} W cm^{-2} in xenon. (a) The continuum–continuum coupling is small, in agreement with numerical evaluations. (b) Electron correlation can be neglected.

Several extensions of this work can be foreseen. First, repeating this type of experiment on an atom like cesium will provide a quantitative test of the theory. This is a necessary step for a definitive understanding of the process. Second, by using high-intensity tunable lasers, a multiphoton photoemission spectroscopy can be initiated.

ACKNOWLEDGMENT

The authors would like to thank C. Manus for many stimulating discussions, particularly on the problem of continuum–continuum coupling.

REFERENCES

Agostini, P., Bonnal, J. F., Mainfray, G., and Manus, C. (1968). *C. R. Hebd. Seances Acad. Sci. Ser. B* **B266**, 790.
Agostini, P., Fabre, F., Mainfray, G., Petite, G., and Rahman, N. K. (1979). *Phys. Rev. Lett.* **42**, 1127.
Agostini, P., Clement, M., Fabre, F., and Petite, G. (1981). *J. Phys. B* **14**, L491.
Aymar, M., and Crance, M. (1981). *J. Phys. B* **14**, 3585.
Ballu, Y. (1980). *Adv. Electron. Electron Phys., Suppl.* **13B**, 257
Boreham, B. W., and Hora, H. (1979). *Phys. Rev. Lett.* **42**, 12, 776.
Crance, M., and Aymar, M. (1980a). *J. Phys. B.* **13**, L421.
Crance, M., and Aymar, M. (1980b) *J. Phys. B.* **13**, 4129–4149.
Dalgarno, A., and Lewis, J. T. (1955). *Proc. R. Soc. London, Ser. A* **A233**, 70.
Fabre, F., Agostini, P., Petite, G., and Clement, M. (1981). *J. Phys. B* **14**, L677.
Fabre, F., Petite, G., Agostini, P., and Clement, M. (1982). *J. Phys. B* **15**, 1353.
Fabre, F., Agostini, P., and Petite, G. (1983). *Phys. Rev. A* **27**, 1682.
Gavrila, M., and Van der Wiel, M. (1978). *Comments Atom. Mol. Phys.* **8**, 1–20.
Gontier, Y., and Trahin, M. (1980). *J. Phys. B* **13**, 4381.
Gontier, Y., Rahman, N. K., and Trahin, M. (1976). *Phys. Rev. A* **14**, 2109.
Hollis, M. J. (1978). *Opt. Commun.* **25**, 395.
Karule, E. (1971). *J. Phys. B* **4**, L67.
Karule, E. (1976). *In* "Multiphoton Process" (J. H. Eberly and P. Lambropoulos, eds.), p. 159. Wiley, New York.

Kibble, T. W. B. (1966). *Phys. Rev. A* **150**, 1060.
Klarsfeld, S. (1970). *Nuovo Cimento Lett.* **3**, 395.
Klarsfeld, S., and Maquet, A. (1979a). *Phys. Lett. A* **73A**, 100.
Klarsfeld, S., and Maquet, A. (1979b). *J. Phys. B* **12**, L553.
Kruit, P., and Read, F. H. (1983). *J. Phys. E: Sci. Instrum.* **16**, 313.
Kruit, P., Kimman, J., and Van der Viel, M. J. (1981). *J. Phys. B* **14**, L597.
Lambropoulos, P. (1976). *Adv. At. Mol. Phys.* **12**, 87.
Martin, E. A, and Mandel, L. (1976) *Appl. Opt.* **15**, 2378.
Morellec, J., Normand, D., and Petite, G. (1982). *Adv. At. Mol. Phys.* **18**, 97.
Nitzan, T., Jortner, J., and Berne, B. (1973). *Mol. Phys.* **26**, 281.
Shiff, L. (1955). "Quantum Mechanics." McGraw-Hill, New York.
Voronov, G. S., and Delone, N. S. (1965). *JETP Lett (Engl. Trans.)* **1**, 42.
Weingartshofer, A., Clarke, E. M., Holmes, J. K., and Jung, Ch. (1979). *Phys. Rev. A* **19**, 2371.
Zernik, W., and Klopfenstein, R. W. (1965). *J. Math. Phys. (N. Y.)* **6**, 262.

7

Multiphoton Free–Free Transitions

A. WEINGARTSHOFER

Department of Physics
St. Francis Xavier University
Antigonish, Nova Scotia
Canada

C. JUNG

Fachbereich Physik
Universität Kaiserslautern
Kaiserslautern, Federal Republic of Germany

I. INTRODUCTION

The topic of this book is multiphoton ionization of atoms. In these processes an electron proceeds from a bound state to a continuum state under the influence of an external electromagnetic field. In the initial state the

155

electron is in a localized state and it needs a supply of energy in order to be removed far away from the rest of the atom. This energy is delivered by the laser field. Also, if the electron already has enough energy to leave the atom, then the external field can interact further with the electron and can deliver additional energy into the system. Hence the electron can proceed from one continuum state into another continuum state after ionization. (Such processes have been studied and are discussed in Chapter 6 of this book.) However, in order to investigate these continuum–continuum transitions, it is not necessary to bring the electron into the initial continuum state by photoionization. In many respects it is simpler to prepare a free electron of the desired energy, let it scatter from a target, and then illuminate the scattering process with a laser. Under these conditions the electron can emit and absorb photons during the scattering process. Of course the initial and final electron scattering states are not completely free states, but only asymptotically free states. Nevertheless, these electron scattering processes under the influence of an external field have been given the name free–free transitions (FF).

It is also possible to consider these processes via the following ideas: During a scattering event, the electron is accelerated and, on the basis of classical electrodynamics, an accelerated charged particle will emit electromagnetic radiation, i.e., bremsstrahlung. For an electron with a kinetic energy of the order of 10 eV, the spontaneous emission of a photon with a perceptible energy has only a negligible probability. And therefore we usually disregard the fact that those processes which we call elastic are, strictly speaking, a sum over all possible bremsstrahlung processes. [For more background information on these problems see Jauch and Rohrlich (1976, Section 16-1).] However, if the electron scattering event takes place inside a laser field, then the induced analog of bremsstrahlung can occur. By the strong laser field, the probability for the emission of photons of the lasing modes is increased significantly so that the emission of laser photons becomes observable. Of course absorption of laser photons also occurs with roughly the same probability. With a sufficiently intense laser field, even the induced emission/absorption of several photons by the same electron has been demonstrated experimentally (see Section III).

Interest in inverse bremsstrahlung or FF photoabsorption has always been strongly motivated for applied reasons: (i) for the one-photon process to account for the infrared opacity of the solar atmosphere; (ii) for multiphoton processes in connection with plasma heating by lasers; (iii) for the development of high-power lasers and the study of stellar interiors, much of this information is needed.

Spontaneous bremsstrahlung was detected in the laboratory by Röntgen in 1895. By contrast, FF photoabsorption grew out of astrophysical observations and the human imagination. Indeed successful calculations of the cross section for the process

$$H + e^- (E_1) + \hbar\omega \to H + e^- (E_2) \qquad (1.1)$$

represent an outstanding triumph in the history of theoretical physics of the 1930s.* The computed results matched very well the quantitative measurements of the opacity in the infrared region of the solar atmosphere. This constituted an elegant confirmation that the 1-photon photoabsorption process summarized in Eq. (1.1) actually occurs in stellar atmospheres. Similar one-photon processes have stimulated considerable theoretical research through to the present (Chandrasekhar and Breen, 1946; Conneely and Geltman, 1981).

Multiphoton processes were investigated later, in the 1960s. The invention and development of high-power lasers provided the necessary impetus in this new direction. Three-photon processes had been observed earlier in the radio-frequency region of the electromagnetic spectrum, but it was the almost instantaneous generation of hot plasmas by focused giant laser pulses on solids and gases that were most intriguing (Morgan, 1975; Keen, 1974). Although it is not always clear how the first few free electrons are produced, the consensus is that in the second stage of the process, the initiatory electrons are accelerated by repeated inverse bremsstrahlung, combined with ionization by an avalanche of collisions with neutral atoms, that lead to the exponential growth in the ion population. The possibility of using laser energy to initiate fusion reactions very soon became the most challenging problem in the physics of laser-produced plasmas. It was discovered that besides inverse bremsstrahlung, which is collisional absorption, there were also collisionless processes or collective effects that become of importance at high laser flux densities (above 10^{14} W/cm^{-2}). However, it is of interest to note that for practical reasons (hot electron preheating), a successful laser fusion experiment has to be performed at laser flux densities such that collisional inverse bremsstrahlung remains of major importance for the coupling of laser energy into the plasma (Key and Hutcheon, 1980).

The description of the applications so far are incomplete. In reality we know that we are always in the presence of laser-induced FF transitions, i.e., electrons exchanging energy with the radiation field, and therefore we have both emission (stimulated bremsstrahlung) and absorption (inverse bremsstrahlung or FF photoabsorption). Schlessinger and Wright (1979) made a quantitative review of the problem of heating plasmas with lasers and find that for plasmas with a Maxwellian distribution of electron energies, energy is absorbed from the laser radiation field: i.e., on the average, energy absorbing encounters outweigh energy emitting encounters. Rosenberg et al. (1980)

---

* The other source of opacity arises from the bound–free photodetachment process: $H^- + \hbar\omega \to H + e^-$.

have succeeded in creating a special distribution of electrons with an inverted population. Under these circumstances, emission outweighs absorption and Rosenberg and co-workers made the interesting observation of the cooling effect of photons in the microwave region. Controlled three-beam experiments (see Section III.B) have shown that for monoenergetic electrons there exists essentially a symmetry between emission and absorption.

The general problem of emission and absorption of hot gases and plasmas has been discussed in the review articles of Gavrila (1977), Gavrila and Van der Wiel (1978), and Geltman (1972, 1977). In this case the FF transitions remain buried in a complex system of atomic processes. Some of the major ones are: collisional excitation; collisional ionization; photoexcitation; photoionization; and in every case, an exact inverse process: de-excitation, recombination, photo-de-excitation or emission, and photorecombination.

It is only quite recently that multiphoton FF transitions have been studied when isolated from other effects. This was done successfully in a controlled three-beam setup (see Section III). These experiments can be regarded as a classical example of laser-modified electron–atom scattering. The technique presents obvious advantages, which have opened up new perspectives for experimental and theoretical investigations. Because of the flexibility of the system, one can see the possibility of studying other processes with the same equipment, i.e., laser-modified Penning ionization has been suggested (George, 1982). It would also be interesting to demonstrate by this method the same cooling effect by photons in the infrared or visible regions, similar to the observations made by Rosenberg et al. (1980) for microwaves.

Among the applications, it should be remarked that much of what has been discussed above provides essential information that is needed for the better understanding of the workings of gas lasers (especially high power) and their development. The optical laser cavity is a rich laboratory where much of laser-modified physics and chemistry takes place. Furthermore, the information is of particular interest to astrophysicists who study plasmas that make up the core of stars, in which there exists a high density of radiant energy.

Free–free transitions are also interesting in themselves because they may perhaps lead to the observation of effects that are, in principle, not observable in radiationless two-body electron–atom scattering. The basic idea was recognized in nuclear physics by Feschbach and Yennie (1962), who proposed the use of bremsstrahlung measurements to obtain information on the off-shell T matrices. The photons involved play the role of a third body and carry away energy and momentum. So the T matrix for the description of the scattering of the remaining massive particles becomes off-shell. Usually off-shell T matrices occur in the description of reactive scattering of several heavy particles, e.g., in dissociative atom–diatomic-molecule scattering the physical on-shell T matrix for the three-body process can be decomposed into a series of two-body off-shell T matrices. But atoms and molecules are complicated structures with many internal degrees of freedom, therefore it is hard to observe in an experiment the effects which are typical for the occurrence of off-shell amplitudes of subsystems. In contrast, the photons in a single-mode laser constitute the theoretically ideal third body. They have no internal structure and they all deliver the same well-defined amount of energy and momentum. It seems likely that FF transitions are the clearest way to create off-shell effects in electron–atom scattering. But, as we shall see below, these off-shell effects do not play any significant role for small photon energies and they have not yet been observed in experiments. Perhaps the development of excimer lasers will change the situation.

Because of the importance of FF transitions in various areas of physics, such processes attracted the attention of theoreticians long ago. The matrix elements that describe one-photon FF transitions in Coulomb scattering have been already calculated by Sommerfeld and Maue (1935). The first approximate treatment of multiphoton FF transitions in a strong laser field was given by Bunkin and Fedorov (1966). For the target they took a local potential model and treated this potential in first-order only.

For a convenient treatment of the problem, one needs an approximation that expresses FF amplitudes by the scattering amplitudes of the corresponding radiationless process. Kroll and Watson (1973) derived a low-frequency approximation (abbreviated lfa) with this property. In the years following, the lfa turned out to be the most useful theoretical description of laser-induced multiphoton FF transitions and has since been generalized in several aspects. In Section II we shall discuss it in detail. Section III contains a discussion of the experimental observation of FF transitions in beam experiments. Section IV gives some final remarks and outlooks and some suggestions for future work in this field. Let us mention here two review articles on free–free transitions which we recommend to the reader: Gavrila and Van der Wiel (1978) and Mittleman (1982).

II. THEORY

A. Motivation for Making a Low-Frequency Approximation

For a completely correct description of FF transitions it would be necessary to take into full account all the interactions between the three systems taking part in the process: the electron, the target (atom or molecule), and the photon beam. This problem is too complicated to be tractable; therefore, let us restrict ourselves to processes where the interaction between laser and target does not play any significant role and can therefore be disregarded, as can be checked. The influence of the laser–target interaction on the process considered here can be estimated by the following considerations: The external field induces an electric dipole moment in the target atom, and this dipole field can be felt by the scattered electrons and can thereby modify the electron–target interaction potential V. Next let us estimate the order of magnitude of this effect. In most of the following discussion we shall consider infrared lasers with a photon energy of the order of 0.1 eV. Because this photon energy is smaller by a factor of 10 to 100 compared to the first excitation energy of the atomic ground states, we use the static polarizability to calculate the induced dipole moment (thereby we overestimate the effect). For atoms the polarizability is between 2×10^{-25} cm^3 (in esu) for a small closed-shell atom like helium and 400×10^{-25} cm^3 for a large alkali atom like cesium.

To see strong signals in free–free transitions we need laser power fluxes in the order of 10^8 W cm^{-2}. For a CO_2 laser this corresponds to an electric field amplitude of 900 esu. Therefore the induced dipole moment is between 1.8×10^{-22} and 3.6×10^{-20} esu for the various atoms. A scattering electron at a distance of 10^{-8} cm will feel a corresponding dipole potential E_d of a magnitude between 10^{-3} and 10^{-1} eV. In electron scattering, dipole effects are usually of importance only at very low energy, where the electron–dipole interaction energy E_d is of the same order of magnitude as the kinetic energy E_i of the electron. In order to see sufficiently strong FF signals, we need incoming kinetic energies of the electron in the order of 10 eV and higher, therefore, it seems to be justified to neglect the laser–atom interaction. This neglect of the laser–target interaction does not apply to molecules that have vibrational transition energies close to the photon energy. In general, the laser frequency must be out of resonance with the target system.

Let us for the moment represent the target by a local potential. There remains the problem of describing the motion of an electron under the simultaneous influence of the laser beam and of the target potential. This problem is still too complicated to be solved exactly and therefore the following two approximation schemes have been applied in the past:

1. It is possible to treat the electron–laser interaction perturbatively and to calculate for any transition amplitude the power series in the electron charge e to any desired order. This procedure is appropriate for low laser powers, as long as only the lowest order gives a significant contribution. It becomes inconvenient as soon as many different orders in e give a contribution of the same order of magnitude. Because we are mainly concerned with high power fields in this chapter, we shall not talk about perturbative methods any longer and consider only nonperturbative approximations of the laser–electron interaction for the rest of the theory section.

2. The complications in FF transitions are caused by the mutual interplay of the electron–laser and electron–target interaction. Therefore it is desirable to find an approximation which decouples these two interactions. The lfa does the job and now we will explain this approximation.

For spontaneous bremsstrahlung, it has been known for a long time (Nordsieck, 1937) that in the low-frequency limit, the cross section factorizes into two terms. One factor is the cross section for the corresponding elastic scattering process. The other factor contains the electron–photon interaction but does not depend on the electron–target interaction. This is exactly the kind of decoupling of the two interactions one needs in order to make the bremsstrahlung process easily tractable. Low (1958) has shown in a quantum-electrodynamical treatment that this factorization is correct in the lowest two orders in the photon energy. Heller (1968) rederived this fact for scattering from a local potential and found the following results, which are interesting for the understanding of the lfa: The contribution to bremsstrahlung, which is lowest order in the photon energy, comes from those graphs in which the electron–photon vertex is outside the electron–target interaction, i.e., in lowest order the photon is attached to the incoming or outgoing electron line. The exact amplitude for this external emission of a photon is expressed in a natural way by an elastic off-shell scattering amplitude. Replacing these off-shell amplitudes by on-shell amplitudes causes a change in the FF amplitude, which is of the next order in the photon energy. The next-order contribution, which shifts the elastic scattering amplitude on the energy shell, is provided exactly by the lowest-order contribution of the amplitude for a photon attached to the internal electron lines. This is an important step, altogether, the two lowest orders of the expansion of the radiative matrix element in powers of the photon energy can be expressed by the radiationless on-shell scattering amplitude taken at shifted momenta. So far everything has been shown for spontaneous one-photon transitions.

Kroll and Watson (1973) have shown that the same results also hold for laser-induced multiphoton processes. The principal result of Kroll and Watson can be given by the following formula:

$$\frac{d\sigma^{\mathrm{ff}}}{d\Omega}(N, \mathbf{p}_f, \mathbf{p}_i) = \frac{p_f}{p_i} J_N^2 \left[\boldsymbol{\alpha} \cdot (\mathbf{p}_f - \mathbf{p}_i)\right] \frac{d\sigma^{\mathrm{el}}}{d\Omega}(\mathbf{p}_f', \mathbf{p}_i') \qquad (2.1)$$

This equation gives the cross section for a process where the electron comes in with momentum $\mathbf{p}_i$ and goes out with momentum $\mathbf{p}_f$, where

$$p_f^2 = p_i^2 + 2mN\hbar\omega \qquad (2.2)$$

$\boldsymbol{\alpha}$ is an abbreviation for $eA\varepsilon/mc\hbar\omega$. The laser field is taken in dipole approximation with the linear polarization ε, m is the electron mass, ω is the laser frequency, J_N is the Bessel function of first kind and order N, A is the amplitude of the vector potential of the field. Equation (2.1) expresses the FF cross section for absorption/emission of N photons by the elastic scattering cross section taken at the shifted momenta

$$\mathbf{p}'_{f/i} = \mathbf{p}_{f/i} - \varepsilon N m \hbar \omega / [\varepsilon \cdot (\mathbf{p}_f - \mathbf{p}_i)] \qquad (2.3)$$

The equality $(\mathbf{p}'_i)^2 = (\mathbf{p}'_f)^2$ shows that the elastic cross section on the right-hand side of Eq. (2.1) is on-shell.

In the meantime Eq. (2.1) has been rederived by various different methods. Let us mention here the particularly straightforward method by Mittleman (1970a). He makes a direct expansion in powers of ω of the free–free T matrix and he also gives explicit expressions for the second-order corrections of the lfa, Eq. (1). Rosenberg (1979a) has given a derivation for arbitrary polarizations of the field and, by not using the dipole approximation, he includes the momentum transfer between laser field and electron.

An analogous formula to Eq. (2.1) holds also for an inclusion of internal states of the target. The following changes are required: in Eq. (2.1) the notation must be extended to allow for a labeling of the initial and final target states, and in Eq. (2.2) the excitation energy of the target must be included into the energy balance (Mittleman, 1980).

Finally, also continuum states of the target have been taken into account, i.e., electron impact ionization of atoms under the additional influence of an external laser field has been considered. In the lfa, the cross section for these processes still factorizes into a laser-dependent factor and into the radiationless electron impact ionization cross section taken at shifted momenta (Banerji and Mittleman, 1981; Cavaliere et al., 1980).

Not much is known about the range of applicability of Eq. (2.1). The following two conditions should be fulfilled in any case:

1. The laser frequency should be sufficiently small so that $\hbar \omega \ll E_i$.
2. The radiationless cross section should be a smooth function of energy.

Condition 1 ensures that photons attached to the external electron lines give the main contribution to the cross section, i.e., condition 1 justifies a separation of the electron–laser and electron–target interaction. Condition 2 will be explained in Section II.D.

Formula (2.1) is also correct, independently of the laser frequency, if the electron–target interaction is treated in first Born approximation on both sides of Eq. (2.1). Of course, in first Born approximation the electron–target interaction is point-like. Then it is self-evident that all electron–laser interaction takes place outside the electron–target interaction and it is obvious that the factorization of the cross section follows. Note that in first Born approximation the momentum shift in the elastic cross section is irrelevant.

B. Derivation of the Kroll–Watson Formula

Because of the importance of Eq. (2.1), let us give a derivation here. To demonstrate the idea we treat the simple case of a local potential and a dipole field. The derivation for more complicated cases can be done along the same pattern.

The Hamiltonian for the motion of an electron under the simultaneous influence of the local potential V and the external dipole field $\mathbf{A}(t) = \boldsymbol{\varepsilon} A \cos \omega t$ is given by

$$H = \frac{\mathbf{p}^2}{2m} - \frac{e\mathbf{A}(t)\mathbf{p}}{mc} + V(\mathbf{x}) \tag{2.4}$$

In dipole approximation the A^2 term does not contain any operator acting upon the electron and therefore it is assumed that this term is already removed by a phase transformation.

For later purposes we need some equations for the elastic T-matrix elements. By $\hat{V}$ we denote the Fourier transform of the potential V. Then the momentum space T-matrix element can be expanded into a power series in $\hat{V}$ as follows:

$$\langle \mathbf{p}_f | T^{\text{el}}(z) | \mathbf{p}_i \rangle = \sum_{n=1}^{\infty} \langle \mathbf{p}_f | T^{(n)\text{el}}(z) | \mathbf{p}_i \rangle \tag{2.5}$$

where

$$\langle \mathbf{p}_f | T^{(n+1)\text{el}}(z) | \mathbf{p}_i \rangle = \int d\mathbf{p}_1 \cdots \int d\mathbf{p}_n \, \hat{V}(\mathbf{p}_f - \mathbf{p}_n) \prod_{j=1}^{n} \hat{V}(\mathbf{p}_j - \mathbf{p}_{j-1}) \left(z - \frac{p_j^2}{2m} \right)^{-1} \tag{2.6}$$

where $\mathbf{p}_0 = \mathbf{p}_i$.

The on-shell T-matrix element for momenta shifted by λ can be expanded into powers of λ and the terms up to first order in λ are

$$\langle \mathbf{p}_f - \lambda | T^{(n+1)\text{el}} \left[\frac{(\mathbf{p}_i - \lambda)^2}{2m} \right] | \mathbf{p}_i - \lambda \rangle$$

$$= \int d\mathbf{p}_1 \cdots \int d\mathbf{p}_n \, \hat{V}(\mathbf{p}_f - \mathbf{p}_n) \prod_{j=1}^{n} \hat{V}(\mathbf{p}_j - \mathbf{p}_{j-1}) \left[\frac{(\mathbf{p}_i - \lambda)^2}{2m} - \frac{(\mathbf{p}_j - \lambda)^2}{2m} \right]^{-1}$$

$$= \int d\mathbf{p}_1 \cdots \int d\mathbf{p}_n \, \hat{V}(\mathbf{p}_f - \mathbf{p}_n) \left[\prod_{j=1}^{n} \hat{V}(\mathbf{p}_j - \mathbf{p}_{j-1}) \times \left(\frac{\mathbf{p}_i^2}{2m} - \frac{\mathbf{p}_j^2}{2m} \right)^{-1} \right]$$

$$\times \left[1 - \sum_{j=1}^{n} \left(\frac{\mathbf{p}_j - \mathbf{p}_i}{m} \right) \cdot \lambda \left(\frac{\mathbf{p}_i^2}{2m} - \frac{\mathbf{p}_j^2}{2m} \right)^{-1} \right] \tag{2.7}$$

The time-dependent wave function for an electron with average momentum $\mathbf{p}$ in the external field $\mathbf{A}(t) = \boldsymbol{\varepsilon}A \cos \omega t$, but without potential V, is given by

$$\Psi(\mathbf{p}|\mathbf{x}, t) = (2\pi\hbar)^{-3/2} \exp[i(\mathbf{p} \cdot \mathbf{x}/\hbar - Et/\hbar + \boldsymbol{\alpha} \cdot \mathbf{p} \sin \omega t)]$$

$$= (2\pi\hbar)^{-3/2} \exp[(i/\hbar)(\mathbf{p} \cdot \mathbf{x} - Et)] \sum_{l=-\infty}^{+\infty} J_l(-\boldsymbol{\alpha}\mathbf{p}) \exp(-il\omega t) \quad (2.8)$$

where $E = p^2/2m$. Its Fourier transform is given by

$$\hat{\Psi}(\mathbf{p}|\mathbf{k}, t) = \delta(\mathbf{p} - \mathbf{k}) \exp[-(i/\hbar)Et] \sum_{l=-\infty}^{+\infty} J_l(-\boldsymbol{\alpha} \cdot \mathbf{p}) \exp(-il\omega t) \quad (2.9)$$

Accordingly, the time-dependent momentum space Green's function for the electron in the field is

$$G(\mathbf{k}_2, t_2|\mathbf{k}_1, t_1) = -\frac{i}{\hbar} \theta(t_2 - t_1) \int d\mathbf{p} \cdot \hat{\Psi}(\mathbf{p}|\mathbf{k}_2, t_2)\hat{\Psi}^*(\mathbf{p}|\mathbf{k}_1, t_1)$$

$$= -\frac{i}{\hbar} \theta(t_2 - t_1) \delta(\mathbf{k}_1 - \mathbf{k}_2) \exp\left[\frac{i}{\hbar}E_1(t_1 - t_2)\right]$$

$$\times \sum_{l,r=-\infty}^{+\infty} J_l(-\boldsymbol{\alpha}\mathbf{k}_2) J_r(\boldsymbol{\alpha}\mathbf{k}_1) \exp(-ir\omega t_1 - il\omega t_2) \quad (2.10)$$

Now we derive the amplitude for a transition from initial state $\Psi(\mathbf{p}_i|\mathbf{x}, t)$ into the final state $\Psi(\mathbf{p}_f|\mathbf{x}, t)$ during a scattering from potential V. We expand the S-matrix element into a power series in V according to

$$S(\mathbf{p}_f, \mathbf{p}_i) = \sum_{n=0}^{+\infty} S^{(n)}(\mathbf{p}_f, \mathbf{p}_i) \quad (2.11)$$

Using the equation

$$\sum_{r=-\infty}^{+\infty} J_{l-r}(x)J_r(y) = J_l(x + y) \quad (2.12)$$

we find

$$S^{(n+1)}(\mathbf{p}_f, \mathbf{p}_i) = \lim_{\substack{t_i \to -\infty \\ t_f \to +\infty}} \int_{t_i}^{t_f} dt_1 \cdots \int_{t_i}^{t_f} dt_{n+1} \int d\mathbf{p}_1 \cdots \int d\mathbf{p}_{n+1}$$

$$\times \int d\mathbf{k}_1 \cdots \int d\mathbf{k}_{n+1} \hat{\Psi}(\mathbf{p}_f|\mathbf{k}_{n+1}, t_{n+1})\left[\prod_{j=2}^{n+1} \hat{V}(\mathbf{k}_j - \mathbf{p}_j)\right.$$

$$\times \left. G(\mathbf{p}_j, t_j|\mathbf{k}_{j-1}, t_{j-1})\right] \hat{V}(\mathbf{k}_1 - \mathbf{p}_1) \cdot \hat{\Psi}^*(\mathbf{p}_i|\mathbf{p}_1, t_1)$$

$$= -2\pi i \sum_{l_1=-\infty}^{+\infty} \cdots \sum_{l_{n+1}=-\infty}^{+\infty} \delta(E_f - E_i - L_{n+1}\hbar\omega)$$

$$\times \int d\mathbf{p}_1 \cdots \int d\mathbf{p}_n \, \hat{V}(\mathbf{p}_f - \mathbf{p}_n) \times J_{l_{n+1}}(\alpha(\mathbf{p}_f - \mathbf{p}_n))$$

$$\times \prod_{j=1}^{n} (E_1 + L_j\hbar\omega - E_j)^{-1} \hat{V}(\mathbf{p}_j - \mathbf{p}_{j-1})J_{l_j}(\alpha(\mathbf{p}_j - \mathbf{p}_{j-1})) \quad (2.13)$$

where $L_j = \sum_{k=1}^{j} l_k$. The $(n+1)$th-order term in V of the T-matrix element for the free–free transition is given by

$$T^{(n+1)\text{ff}}(N, \mathbf{p}_f, \mathbf{p}_i)$$

$$= \sum_{l_1=-\infty}^{+\infty} \cdots \sum_{l_n=-\infty}^{+\infty} \int d\mathbf{p}_1 \cdots \int d\mathbf{p}_n \, \hat{V}(\mathbf{p}_f - \mathbf{p}_n)J_{l_{n+1}}(\alpha(\mathbf{p}_f - \mathbf{p}_n))$$

$$\times \prod_{j=1}^{n} (E_i + L_j\hbar\omega - E_j)^{-1} \cdot \hat{V}(\mathbf{p}_j - \mathbf{p}_{j-1})J_{l_j}(\alpha(\mathbf{p}_j - \mathbf{p}_{j-1})) \quad (2.14)$$

where $l_{n+1} = N - L_n$.

Next we expand all denominators in powers of ω and keep terms up to the first order

$$(E_i + \hbar\omega L_j - E_j)^{-1} = \frac{1}{E_i - E_j} - \frac{L_j\hbar\omega}{(E_i - E_j)^2}$$

Inserting into Eq. (2.14) and collecting terms of order 0 gives, after applying the addition theorem Eq. (2.12) l times

$$T^{(n+1)\text{ff}}_{(0)}(N, \mathbf{p}_f, \mathbf{p}_i) = J_N(\alpha(\mathbf{p}_f - \mathbf{p}_i)) \int d\mathbf{p}_1 \cdots \int d\mathbf{p}_n \, \hat{V}(\mathbf{p}_f - \mathbf{p}_n)$$

$$\times \prod_{j=1}^{n} \hat{V}(\mathbf{p}_j - \mathbf{p}_{j-1}) \cdot (E_i - E_j)^{-1} \quad (2.15)$$

For the contributions of first order in ω we find

$$T^{(n+1)\text{ff}}_{(1)}(N, \mathbf{p}_f, \mathbf{p}_i) = -\sum_{j=1}^{n} \sum_{L_1=-\infty}^{+\infty} \cdots \sum_{L_n=-\infty}^{+\infty} \hat{V}(\mathbf{p}_f - \mathbf{p}_n) \cdot J_{N-L_n}(\alpha(\mathbf{p}_f - \mathbf{p}_n))$$

$$\times \frac{L_j\hbar\omega}{E_i - E_j} \prod_{k=1}^{n} \hat{V}(\mathbf{p}_k - \mathbf{p}_{k-1})$$

$$\times J_{L_k - L_{k-1}}(\alpha(\mathbf{p}_k - \mathbf{p}_{k-1}))(E_i - E_k)^{-1}.$$

For any fixed j we use addition theorem (2.12) up to the L_{j-1} summation and use the addition theorem

$$\sum_{r=-\infty}^{+\infty} rJ_r(y)J_{N-r}(x) = \frac{Ny}{x + y} J_N(x + y) \quad (2.16)$$

for the remaining summations and obtain

$$T_{(1)}^{(n+1)^{\mathrm{ff}}}(N, \mathbf{p_f}, \mathbf{p_i})$$

$$= -\sum_{j=1}^{n} \int d\mathbf{p_1} \cdots \int d\mathbf{p_n}\, \hat{V}(\mathbf{p_f} - \mathbf{p_n}) \frac{N\hbar\omega}{E_i - E_j} \cdot \frac{\alpha(\mathbf{p}_j - \mathbf{p_i})}{\alpha(\mathbf{p_f} - \mathbf{p_i})} J_N(\alpha(\mathbf{p_f} - \mathbf{p_i}))$$

$$\times \prod_{k=1}^{n} \hat{V}(\mathbf{p}_k - \mathbf{p}_{k-1})(E_i - E_k)^{-1}. \tag{2.17}$$

Comparison with Eq. (2.7) and summation over all n gives

$$(T_{(0)}^{\mathrm{ff}} + T_{(1)}^{\mathrm{ff}})(N, \mathbf{p_f}, \mathbf{p_i})$$

$$= J_N(\alpha(\mathbf{p_f} - \mathbf{p_i}) \cdot \langle \mathbf{p_f} - \lambda | T^{\mathrm{el}}[(\mathbf{p_i} - \lambda)^2/2m] | \mathbf{p_i} - \lambda \rangle \tag{2.18}$$

where

$$\lambda = N\hbar\omega m\alpha/\alpha(\mathbf{p_f} - \mathbf{p_i}).$$

Equation (2.18) leads directly to Eq. (2.1).

C. Averaging for an Inhomogeneous Laser Beam

In Eq. (2.1) the power density of the laser has been assumed to be independent of space and time. This assumption is never fulfilled for any real laser and therefore it is necessary to consider the effects of an intensity variation of the field before it is possible to apply Eq. (2.1) to an experiment. In the following we denote the space-and-time dependent power flux of the laser field by $F(\mathbf{x}, t)$. It is reasonable to describe the laser beam by a modulated plane wave with a slowly varying modulation function. In this section it is more convenient to write the argument of the Bessel function in Eq. (2.1) in terms of the power flux F. We set $\alpha \cdot (\mathbf{p_f} - \mathbf{p_i}) = \beta\sqrt{F}$, where $\beta = e\boldsymbol{\varepsilon} \cdot (\mathbf{p_f} - \mathbf{p_i})\sqrt{8\pi c}/mch\omega^2$.

Next we must consider the number of events which a detector counts in an experiment, performed with a space-and-time dependent field intensity. We shall assume an adiabatic variation of the intensity because of the following reasons: To be specific, let us assume that the experiment is done with a CO_2 laser like the one used in the experiments of Weingartshofer et al. (1977). The frequency separation between adjacent modes in a CO_2 laser is in the order of 10^7 to 10^8 s^{-1} (the exact value depends on the size of the resonator), and normally not very many modes contribute to the beam. Therefore the bandwidth $\Delta\omega$ is at most in the order of 10^9 s^{-1}. Accordingly, the time over which the intensity varies significantly is at least in the order of 10^{-9} s. This is long compared to the duration of an electron–atom (molecule) scattering process, even in the case of resonances. In the case of other laser types, which allow faster intensity fluctuations, it would be necessary to exclude the occurrence of sharp electron–target scattering resonances. In addition, the effective range of the electron–atom (molecule) potential is small compared to the laser wavelength and therefore also small compared to the spatial intensity variations. This means that the electron–target scattering event does not feel any effect caused by the intensity variations of the laser field. Therefore we describe each single FF process by Eq. (2.1), where F in the argument of the Bessel function is the flux value at the particular space–time point at which the process occurs. Each space–time point $(\mathbf{x}, t)$ is to be weighted by $w(\mathbf{x}, t)$, the density of electron–target collisions at this point. The detector records events occurring inside a macroscopic space–time volume, and therefore the total number of observed events is proportional to Eq. (2.1) multiplied by the weight function $w(\mathbf{x}, t)$ and integrated over the observed space-time volume VT. Accordingly, the directly observable quantities in FF experiments are the numbers

$$R_N = \int_V d\mathbf{x} \int_T dt\, w(\mathbf{x}, t) J_N^2(\beta\sqrt{F(\mathbf{x}, t)}) \qquad (2.19)$$

Depending on the apparatus, $w(\mathbf{x}, t)$ may contain a factor describing the different efficiency of the detector for events occurring at different space–time points. It is useful to normalize $w(\mathbf{x}, t)$ so that

$$\int_V d\mathbf{x} \int_T dt\, w(\mathbf{x}, t) = 1 \qquad (2.20)$$

The quantities R_N are the relative ratios with which the various final electron energies $E_f(N) = E_i + N\hbar\omega$ are observed in an experiment.

Bessel functions have the properties (see Eq. 8.536 in Gradshteyn and Ryzhik, 1965).

$$\sum_{N=-\infty}^{+\infty} J_N^2(z) = 1 \qquad (2.21)$$

and

$$\sum_{n=0}^{\infty} (2N + 2n)(2N + n - 1)!\, \frac{1}{n!}\, J_{N+n}^2(z) = \frac{(2N)!}{(N!)^2} \left(\frac{z}{2}\right)^{2N} \qquad (2.22)$$

for $N = 1, 2, \ldots$. This leads to the following sequence of sum rules for free–free transitions:

$$\sum_{N=-\infty}^{+\infty} R_N = 1 \qquad (2.23)$$

and

$$\frac{(N!)^2}{(2N)!} \frac{2^{2N}}{\beta^{2N}} \sum_{n=0}^{\infty} (2N + 2n)(2N + n - 1)!\, \frac{1}{n!}\, R_{N+n}$$

$$= \int d^3x \int dt\, w(\mathbf{x}, t) F^N(\mathbf{x}, t) \qquad (2.24)$$

for $N = 1, 2, \ldots$. Equation (2.23) is the sum rule, which says that the total number of electrons summed up over all final energies, which go into a given direction, is independent of the properties of the laser beam, and therefore it is the same as in the absence of the laser. The sum rule (2.23) has been discussed extensively by Jung (1979) and by Rosenberg (1979b).

The particular case of Eq. (2.24) for $N = 1$ has been discussed by Jung (1980). It implies that the average energy transfer from the laser beam onto the electrons does not depend on the exact power distribution of the laser beam in space and time, but only on the average power in the scattering volume. This average power is given by the right-hand side of Eq. (2.24) for $N = 1$. Jung (1981) has suggested that the problem can be turned around so that (2.24) can be used to extract properties of the laser out of measured FF cross sections.

A different approach to the averaging of the cross section over the temporal laser flux variation has been used by Zoller (1980). He describes the laser field by a chaotic field representing a large number of uncorrelated modes. This corresponds to a field undergoing Gaussian amplitude fluctuations. His final result consists in replacing the squared Bessel function in (2.1) by $e^{-\zeta^2/2} I_N(\tfrac{1}{2}\zeta^2)$, where

$$\zeta = 2e(\mathbf{p}_f - \mathbf{p}_i) \cdot \boldsymbol{\varepsilon}(\langle |E(t)|^2 \rangle)^{1/2} / mh\omega$$

and I_N is the Bessel function of order N with imaginary argument, $\langle |E(t)|^2 \rangle$ is the correlation function of the time-dependent field amplitude $E(t)$.

For a small field intensity (i.e., $\zeta \ll N$) the cross section in the chaotic field is larger by a factor $N!$ than the cross section in a single-mode field of the same total power. For high field intensities (i.e., $\zeta \gg N$) the oscillations of the Bessel functions are averaged out.

The averaging over a spatially Gaussian intensity variation has been presented by Daniele et al. (1980).

D. Inclusion of Electron–Target Resonances

The incoming electron can emit/absorb photons and thereby change its energy. It can probe the target potential at various energies. Nevertheless, Eq. (2.1) contains the radiationless electron–target cross section only at one particular energy. This restriction to one value of the energy is justified only if the radiationless cross section does not change substantially within an energy interval of several photon energies. Therefore condition 2 given at the end of Section II.A is required.

Condition 2 is not necessarily fulfilled for electron–atom scattering because of the occurrence of sharp resonances. Therefore Eq. (2.1) cannot be used to investigate how resonances of radiationless electron–atom scattering show up in FF transitions.

In order to circumvent this problem we could try to abandon the lfa completely and start with an exact formulation of the scattering process inside the laser field. Dyachkov *et al.* (1975) have investigated resonances in 1-photon FF transitions without using the lfa. But it is difficult to extract general statements out of the resulting equations, and it is not obvious how to generalize the method to the case of multiphoton processes in strong fields. Krüger and Jung (1978) have given a version of the lfa for FF transitions which includes the effects of sharp resonances in the radiationless electron-target scattering. Let us briefly mention the essential ideas of this approach and quote the main results in the next paragraph.

The development of a FF transition can be split into three stages. In the first stage the electron comes in from the electron gun with energy E_i and enters the laser beam where it can emit/absorb photons and thereby accept any energy given by $E_k = E_i + k\hbar\omega$, where k is an integer. It follows from Eq. (2.9) that the amplitude for the electron to attain the energy E_k is given by the Bessel function $J_k(-\boldsymbol{\alpha} \cdot \mathbf{p}_i)$. In the second stage, which describes the interaction of the electron with the target, the essential approximation consists in describing this interaction by the corresponding radiationless scattering amplitude taken at the respective value of the intermediate electron energy E_k. Thereby we neglect the laser-electron interaction during the electron-target scattering. This approximation is supposed to be good for a sufficiently small photon energy. In the third stage the electron leaves the target interaction region and can again emit/absorb photons. The amplitude for emission/absorption of $N - k$ photons is given by $J_{N-k}(\boldsymbol{\alpha} \cdot \mathbf{p}_f)$.

Finally the electron leaves the laser beam and arrives at the detector with energy $E_f = E_i + N\hbar\omega$. Accordingly, the amplitude to proceed from the initial state to the final state via a particular intermediate electron energy is given by the product of the three amplitudes mentioned above. For a schematic representation of the three stages, see Weingartshofer *et al.* (1981). The total amplitude for the transition is obtained by summation over all possible intermediate energies. The result gives the amplitude for a FF transition in which the electron comes in with energy E_i, goes out with energy $E_f = E_i + N\hbar\omega$ and is scattered at an angle θ.

$$f^{\mathrm{ff}}(E_i, \theta, N) = \sum_k J_{N-k}(\boldsymbol{\alpha} \cdot \mathbf{p}_f) f^{\mathrm{el}}(E_i + k\hbar\omega, \theta) J_k(-\boldsymbol{\alpha} \cdot \mathbf{p}_i) \qquad (2.25)$$

An analogous formula holds for FF transitions with excitation of the target, if we take, on the right-hand side, the corresponding scattering amplitude for radiationless excitation of the target and take the excitation energy into the energy balance.

Equation (2.25) is correct in lowest order in ω. Corrections in the next two orders in ω have been given by Mittleman (1979b). Another quite elegant derivation correct in the two leading orders in ω and valid for all polarizations of the laser has been given by Rosenberg (1981). The essential modification of taking into account the first-order corrections is a momentum shift in the radiationless scattering amplitude similar to the one in Eq. (2.1).

Equation (2.25) has been used by Jung and Taylor (1981) to show that FF transitions can be used to filter out resonances in an electron scattering process. Assume that f^{el} has an isolated resonance at energy E_R whose width Γ is small compared to $\hbar\omega$. Separate f^{el} into a resonance and a background part

$$f^{el}(E, \theta) = f_R(E, \theta) + f_{BG}(E, \theta) \qquad (2.26)$$

f_{BG} is supposed to be nearly constant as function of energy over an interval whose length is large compared to $\hbar\omega$. Therefore we pull f_{BG} out of the k-sum in (2.25), apply Eq. (2.12), and obtain

$$f^{ff}(E_i, \theta, N) = \sum_k J_{N-k}(\boldsymbol{\alpha} \cdot \mathbf{p}_f) f_R(E_i + k\hbar\omega, \theta) J_k(-\boldsymbol{\alpha} \cdot \mathbf{p}_i)$$

$$+ J_N[\boldsymbol{\alpha} \cdot (\mathbf{p}_f - \mathbf{p}_i)] f_{BG}(E_i, \theta) \qquad (2.27)$$

Because of $\Gamma \ll \hbar\omega$, f_R is essentially different from zero only for the one particular k-value for which $E_i + k\hbar\omega \approx E_R$, i.e., at most one term in the k-sum in Eq. (2.27) is of importance for a given value of E_i. If the incoming electron energy E_i is scanned and the detector acceptance energy E_f is changed simultaneously so that $E_f = E_i + N\hbar\omega$ always, then the detector shows a sequence of resonance structures at the energies $E_i = E_R - k\hbar\omega$, where k is an integer. The width of all these resonances is Γ. Of course, for a given laser intensity, only a few peaks are strong enough to be seen. The strength of the various resonance structures and their interference with the background depends on the laser intensity and on the scattering geometry.

If $\boldsymbol{\alpha}$ is perpendicular to $\mathbf{p}_f - \mathbf{p}_i$ then the argument of the Bessel function in the background term in Eq. (2.27) is zero and the Bessel function itself is $J_N(0) = \delta_{N,0}$. This leads to the interesting prediction that for some special choices of the scattering geometry there is no background at all in photon emission and absorption processes. Therefore FF transitions might be used to measure resonances of the electron–target scattering without the influence of any background effects. In order to obtain a strong resonance signal, an experimentalist is advised to bring $\mathbf{p}_i$ and $\mathbf{p}_f$ as parallel to each other as the apparatus allows and bring $\boldsymbol{\varepsilon}$ in the direction between $\mathbf{p}_f$ and $\mathbf{p}_i$. Then $\boldsymbol{\alpha} \cdot \mathbf{p}_i$ and $\boldsymbol{\alpha} \cdot \mathbf{p}_f$ are both as large as possible and $\boldsymbol{\alpha} \cdot (\mathbf{p}_f - \mathbf{p}_i) \approx 0$ for the case of elastic electron–target interaction. For a more extensive discussion of these considerations see Jung and Taylor (1981). These considerations have been derived from the lfa Eq. (2.27), but they also have been confirmed by model calculations not using the lfa.

E. Some Remarks on the Use of the Low-Frequency Approximation

So far the lfa is well understood. But it is not yet clear if Eq. (2.1) is also valid for long-range electron–target interactions of the Coulomb plus short-range potential type. Rosenberg (1979c) claims that Eq. (2.1) holds also in this case. This has been disputed by Banerji and Mittleman (1982). The reason for this discrepancy is not yet clear.

In the lfa we express FF transition amplitudes by radiationless scattering amplitude and by some special functions of mathematical physics like, e.g., Bessel functions. Therefore such processes which are described completely by the lfa do not provide any new information on electron–target scattering. In order to gain information on radiationless scattering cross sections we do not need to perform a FF experiment. New physics will become observable only under those circumstances where the lfa breaks down and off-shell effects may play a significant role. These off-shell corrections are partly contained in the second-order corrections calculated by Mittleman (1979a,b). These second-order corrections are quite complicated in form, and it is difficult to derive general statements about observable effects caused by them.

Coulter and Ritchie (1981) have investigated one-photon FF transitions with spin-dependent electron–target interaction. They have looked at spin polarization of the scattered electrons. Spin polarization is not contained in the lfa. Only the off-shell corrections give an analog to the Fano effect in bound–free transitions. Therefore the observations of any spin polarization of the outgoing electrons in FF transitions with an unpolarized incoming electron beam would be a clear indication for off-shell contributions.

One particular type of FF transitions, to which the lfa can never be applied are resonance–resonance transitions. Langendam *et al.* (1978) have attempted this kind of experiment. In these processes the electron comes in exactly at a resonance energy and undergoes an electromagnetic transition to another resonance state.

Therefore it is essential to the formation of the process that the target potential acts upon the electron before and after the emission/absorption of the photon. But just these contributions to the amplitude which are caused by internal emission/absorption are not described properly by the lfa. For a theoretical treatment and some model calculations of 1-photon resonance–resonance transitions see Jung and Krüger (1979).

F. Numerical Computations

As far as we know, nearly all numerical computations for FF transitions that have been done so far treat the electron field interaction perturbatively. [One exception is a model calculation by Jung and Taylor

(1981). But this model is one-dimensional and therefore uses a somewhat unrealistic electron–target model.] There are a lot of first-order computations for FF transitions on particular atomic and molecular targets. For reference see the extensive bibliography by Gallagher (1977). Let us mention here briefly a few more recent papers which are relevant for the numerical tests of the lfa. Comparisions of calculations with and without lfa using multichannel square well potential models have been reported by Krüger and Schulz (1976), Jung and Krüger (1978), and Shakeshaft and Robinson (1982). These computations show that the lfa describes all features qualitatively correctly and is quantitatively good for sufficiently low laser frequencies. This holds also for the case, that one electron–target resonance is involved in the process.

A computation of resonant FF transitions using a realistic description of a hydrogen target has been published by Conneely and Geltman (1981). Unfortunately, in this paper numerically exact computations involving resonances have been compared to the lfa given in Eq. (2.1) and not to the resonance version given in Eq. (2.25). Therefore, it is not possible to extract statements about the quality of the resonant lfa.

III. EXPERIMENT

A. General Remarks

The idea of measuring FF transitions originated in conversations that one of the authors (A.W.) had with D. Andrick and H. Krüger in the laboratory of Professor H. Ehrhardt at the University of Freiburg (now at the University of Kaiserslautern). Andrick measured the one-photon FF process with a 50-W cw CO_2 laser (Andrick and Langhans, 1976, 1977). The success of the experiment was founded on the theoretical work of Krüger who calculated cross sections for the one-photon process in first-order perturbation theory in the radiation field (Krüger and Schulz, 1976). Although there were other theoretical papers at the time (e.g., Section II), Krüger provided the guidelines for the experimental work by pointing out two important conclusions: (1) in the above approximation the FF cross section is proportional to the fourth power of the wavelength of the laser and this fact determined the choice of the CO_2 laser, and (2) the numerical results for the cross section clearly indicated that the experiment was, indeed, feasible with existing technology in conventional electron spectroscopy.

The idea of measuring multiphoton FF transitions grew out of a Canadian invention, the TEA CO_2 laser of Beaulieu (1970). The two basic conclusions of the theoretical work of Krüger made this high-power laser the ideal choice. Strictly speaking, of course, we have to make use of the

Kroll–Watson formula (2.1), which shows that in the case of multiphoton FF transitions the dependency on the flux density F, the wavelength λ of the laser (or the frequency ω), and the incoming electron energy E_i are respectively contained in the argument of the Bessel function as $\sqrt{F}$, λ^2, $(1/\omega^2)$, and $\sqrt{E_i}$ ($E = p^2/2m$). It is instructive to compare the required flux densities for two different lasers, e.g., the CO_2 TEA and the Nd^{3+}-glass laser, to attain the same value for the argument of the Bessel function in the Kroll–Watson formula. A conservative value would be in the order of 2.5 for an electron energy $E_i = 10$ eV. The required flux densities are:

$$CO_2 \text{ TEA laser } (\hbar\omega = 0.117 \text{ eV}): \quad F = 10^8 \text{ W cm}^{-2}$$

$$Nd^{3+}\text{-glass laser } (\hbar\omega = 1.17 \text{ eV}): \quad F = 10^{12} \text{ W cm}^{-2}$$

This comparison shows the advantage of using a CO_2 laser. One apparent disadvantage is that the energy resolution of the electron spectrometer has to be narrow since the CO_2 laser photon energy is ten times smaller than the Nd^{3+}-glass laser photon energy. Experimentally however this presents no serious problems, as the results of the measurements show. Eventually this should be of no disadvantage at all if one considers that the future interest in FF measurements lies in the investigation of effects in the vicinity of sharp resonances and where a good energy resolution will be highly desirable. The small photon energy of the CO_2 laser also represents an enormous advantage in connection with the new data acquisition system as explained below.

In conjunction with good resolution are low electron counting rates. The longer CO_2 laser pulse, lasting several microseconds, is another advantage because it provides a longer sampling time period per shot.

B. Description of the Controlled Three-Beam Apparatus

The experimental arrangement is represented in Fig. 1, showing the geometrical configuration of the three beams: laser, electron, and atom beam (which is perpendicular to the scattering plane defined by the incoming $\mathbf{p}_i$ and detected outgoing $\mathbf{p}_f$ electron beams). The apparatus is essentially an electron spectrometer that has been described in the literature (Weingartshofer et al., 1974) and is provided with a rock salt window to bring in the laser beam. The CO_2 laser (model: Lumonics TEA-103-1) has a photon energy of 117 meV and an energy output of 15 J per pulse. It is operated in multimode optical configuration and the temporal shape of a typical pulse is shown in Fig. 2. Notice that the fast oscillations of the intensity (shown on the oscilloscope trace, Fig. 2b) are averaged out in the figure. The laser beam is polarized with the polarization vector ε parallel to the scattering plane

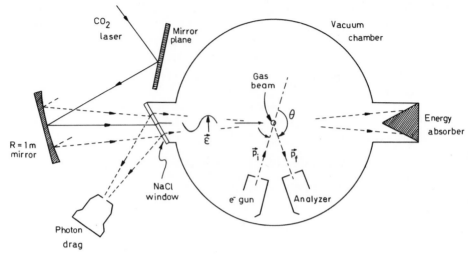

Fig. 1. Geometry of the controlled three-beam apparatus for the measurement of multiphoton free–free cross sections. Gas beam is perpendicular to scattering plane.

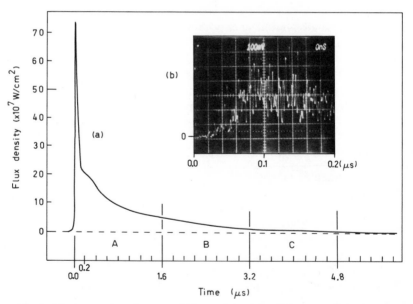

Fig. 2. CO_2 laser pulse. (a) Temporal shape of the pulse. Time intervals A, B, and C are electronically selected by the fast counter described in the text. (b) Oscilloscope trace of the first 200 ns of the laser pulse. Flux density is given for focused beam (4 mm^2) at scattering center.

of the electrons (p_i and p_f). At the moment it is impossible to make it completely parallel because of the multiple reflections on the mirrors, in particular two plane mirrors of a periscope that is not shown in Fig. 1. The beam is focused into the scattering center with a gold mirror ($R = 1$ m) and forms an angle ψ with p_i. The gas beam, perpendicular to the scattering plane, is produced by effusion from a hypodermic needle (inside diameter = 0.5 mm).

C. Experimental Results

Experimental results of multiphoton FF transitions have been reported (Weingartshofer *et al.*, 1977, 1979, 1981, 1982) and the reader is referred to these papers for details. We want to present here a summary of the experimental results obtained to date and discuss briefly the projected modifications of the instruments and the improvement in quality of FF spectra that we hope to obtain.

Until now all the observations have been done on elastically scattered electrons, i.e., incoming and final electron energies are equal ($E_i = E_f$) in the absence of the laser. In the presence of the laser we observe laser-modified elastic scattering, where the scattered electrons can change their kinetic energy to $E_f = E_i \pm N\hbar\omega$ (where N is an integer). As many as eleven photons have been observed to interact with the electrons. Although other gases have been tried (H_2 for instance, see Weingartshofer *et al.*, 1979) most of the experiments so far have been carried out with argon as a target. It has a large elastic cross section in the backward and in the forward directions for an electron energy E_i of about 11 eV (Weingartshofer *et al.*, 1974). The experiments are still in the exploratory stage. It has always been desirable to work under optimum conditions to observe strong FF effects in order to extract quantitative information that can be better compared with the theory. The Kroll–Watson formula [Eq. (2.1)] indicates that two experimental conditions are required to observe a strong FF effect: a large scattering cross section for a sufficiently high value for E_i and the largest possible value that the geometry of the instrument will allow in order to maximize the scalar product SP $= \boldsymbol{\varepsilon} \cdot (p_f - p_i)/2p$. Therefore most of the measurements have been carried out at a scattering angle of $\theta = 154°$. Nevertheless, some measurements have been done under different conditions: measurements in the forward direction and with the laser polarization vector $\boldsymbol{\varepsilon}$ perpendicular to the scattering plane (Weingartshofer *et al.*, 1983).

The first observed multiphoton FF transitions (Weingartshofer *et al.*, 1977) are shown in Fig. 3. The FF spectrum was measured for $E_i = 11.00$ eV and $\theta = 154°$. The experimental points in the presence of the laser (curve A)

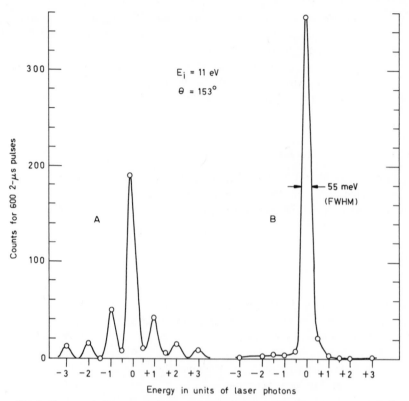

Fig. 3. Energy gain/loss spectra for e^-–Ar scattering (A) in the presence of the laser field and (B) without the laser field. The circles show the measured experimental points and the estimated process is drawn with a solid line. The energy resolution of the electron-gun–detector combination is 55 meV [full width at half maximum (FWHM)]. [From Weingartshofer *et al.* (1977).]

and some hundred microseconds later in the absence of the laser (curve B) were always measured consecutively for each detected electron energy E_f to ensure identical experimental conditions and valid comparisons. Since the laser is pulsed, the scattered electron intensity was collected for several hundred laser pulses (600 in this case) and is plotted along the ordinate. The abscissa gives the energy of the detected electrons in units of laser photons, where 0 photons correspond to an electron energy $E_f = E_i = 11.00$ eV. The other energies are $E_f = E_i \pm k\hbar\omega$, where $k = \frac{1}{2}, 1, \frac{3}{2}, 2$, and 3. The fact that the counts drop practically to zero for half-multiples of a photon is a clear confirmation that the one-photon and two-photon processes are distinctly resolved and represents a direct verification that we are observing, indeed, quantum effects. The measurements were recorded with two gated counting systems, A and B, which were started by the photon drag shown in Fig. 1. Counter A opens its gate after a very small delay of 150 ns (travel time of the electrons from the scattering center to the detector) and remains open for 2 μs, thereby detecting the scattered electrons that have interacted with the radiation field, more exactly the first 2-μs portion of the laser pulse (see Fig. 2). Counter B opens its gate only after a long delay of 100 μs, i.e., well after the laser pulse has disappeared, and therefore measures the elastically scattered electrons in the absence of the field but otherwise identical experimental conditions as the point on curve A for the same energy E_f. In order to improve on the statistics of the measurements, the gate in B remains open for a period of 200 μs but the reported number in Fig. 3 is only for a 2-μs time interval.

In these measurements the energy resolution of the spectrometer was 55–60 meV (full width at half height). This agrees with the measured width of the elastic peak and is indicated with arrows in Fig. 3 (curve B). Figure 3 (curve A) shows clearly a redistribution in energy of the original mono-energetic electrons but it can be observed that the sum of the peak counts shows that there is definite tendency to comply with the sum rule [Eq. (2.23)]. The experimental points have been joined with a solid line to give a more realistic energy distribution of all the electrons within the energy width of 55 meV since they are not strictly monoenergetic. To test for the sum rule, the areas under the curves A and B should be compared rather than the counts at the height of the peaks.

The symmetry of the FF spectrum about the 0-photon peak is an indication that the number of absorbed photons is the same as the number of emitted photons in the case of a monoenergetic beam of electrons confirming the Kroll–Watson formula. It is interesting to consider FF transitions for other electron energy distributions [e.g., Schlessinger and Wright, (1979) and Rosenberg et al. (1980)]. Some pertinent remarks to this effect were made in Section I.

Consideration of the temporal shape of the laser pulse in Fig. 2 reveals that in the previous experiment only the first 2 μs portion containing the large spike was put to use. It seemed, therefore, highly desirable to make full use of the pulse and at the same time establish the dependency of the FF cross section on the laser flux density. With this aim in mind, a fast digital counter was constructed having four digital display channels. The first three designated A, B, and C sample equal and consecutive time slices (1.2 μs each) of the scattered electrons in the presence of the laser pulse as shown in Fig. 2. The fourth channel D was delayed for a time long enough to ensure that the detector was counting scattered electrons without the laser field (same function as counter B in the previous experiment). The results obtained with this counter are shown in Fig. 4. It applies to e$^-$–Ar scattering in the

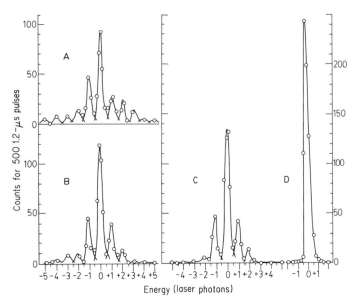

Fig. 4. Energy gain/loss spectra for e^-–Ar scattering showing dependence on laser flux. Spectra A, B, and C show count rates of electrons scattered in time intervals A, B, and C of laser pulse in Fig. 2. Spectrum D measured without the laser. Experimental points ($\bigcirc$) were measured at fractional and integral photon energies to determine the outline of the process. ($\theta = 155°$ and $E_i = 11.72$ eV.) See text for additional information. [From Weingartshofer *et al.* (1979).]

backward direction ($\theta = 155°$ and $E_i = 11.72$ eV). Here we compare the elastic cross section (channel D, without laser) with FF cross sections as a function of a decreasing flux density corresponding to the pulse time slices shown in Fig. 2 and recorded in channels A, B, and C. We searched for scattered electrons at energies given by $E_f = 11.72 \pm k\hbar\omega$, where k is a fractional—and integral—number. Although the statistical errors are high, it is remarkable to observe the experimental points following very nicely the outline of the envelope which was simply assumed in the previous spectrum, Fig. 3. Spectra A, B, and C show a distinct dependency on a decreasing laser flux density. The symmetry (absorption versus emission) observed in A is striking in comparison with B and C, which are very symmetrical. However the statistical errors are too high at the moment to consider this as conclusive. One has to keep in mind that it takes up to half an hour to collect data for one energy E_f and anywhere between 15 to 24 hr to record a complete spectrum. The interesting point to be made here is that recent developments in detection systems can drastically reduce the time scale while improving the quality of the spectra at the same time. This will be further discussed at the end of this section.

The spectra in Figs. 3 and 4 clearly demonstrate the emission and absorption of photons in an elementary FF process. The next step in the development was an attempt to improve experimental conditions and put the emphasis on quantitative measurements that would allow for a better comparison with existing theory. With this in mind, the energy resolution of the spectrometer was reduced to better than 40 meV as compared to 55–60 meV for the previous work. There is a time-saving advantage in this since the spectra may now be measured by simply detecting the electrons that have changed their energies by multiples and half-multiples of a photon energy and omitting smaller fractions. To a good approximation, the number of electrons that have absorbed/emitted a number of photons is now more representative in the peak height. Another improvement consisted in the reduction of the size of the focused laser beam in the scattering center to 4 mm^2 by using the combination of a spherical copper mirror ($f = +150$ cm) and a NaCl plano–convex lens ($f = +15$ cm). In earlier measurements the focal region was 10 mm^2. The cross section of the focused electron beam is in the order of 4 mm^2. Extreme precautions were taken by frequently monitoring the overall alignment, electron energy E_i, and noise pulses for those unwanted. With these modifications a series of FF spectra were measured for the following parameters: four incident electron energies in the backward direction, two incident electron energies in the forward direction, and one measurement by turning the polarization vector perpendicular to the scattering plane. The data were collected with the previously described fast digital counter having the four digital display channels A, B, C, and D. For complete details of this series of quantitative measurements, see Weingartshofer et al. (1982). We show in Figs. 5 and 6 one example of the spectra in this series ($E_i = 15.80$ eV and $\theta = 155°$). Data were collected for 500 1.6-μs slices of the laser pulse (see Fig. 2). Figure 5 shows the results of channel A and Fig. 6 the results of channels B and C. There is no particular interest in showing channel D; however, the accumulated counts in this channel (elastic cross section) were 436. This number appears in the upper right-hand side of each spectrum in the following ratios:

Channel A: $\dfrac{431}{436} = 0.95 \pm 0.08$

Channel B: $\dfrac{447}{436} = 1.03 \pm 0.08$

Channel C: $\dfrac{502}{436} = 1.07 \pm 0.07$

The ratios represent the sums of the R_N (22 peaks) that were discussed in Section II.C and which are directly observable quantities and measured from

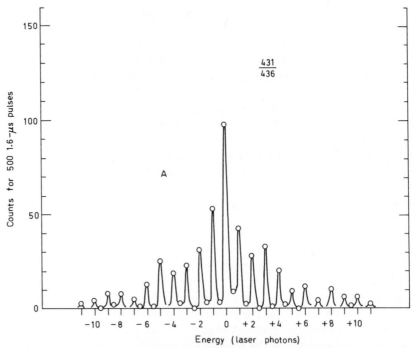

Fig. 5. Multiphoton FF electron energy spectrum for e^-–Ar scattering measured in the presence of the flux density contained in the time slice A of the laser pulse shown in Fig. 2. ($\theta = 155°$ and $E_i = 15.80$ eV.) The electron energy resolution was better than 40 meV and the laser pulse was focused to 4 mm^2 in the target region. Ratio is an indication of the validity of the sum rule. See text for additional information. [Reprinted with permission from Weingartshofer *et al.* (1983). Copyright 1983 by The Institute of Physics.]

the spectra (the quantities R_N are the relative ratios with which the various final electron energies $E_f = E_i + N\hbar\omega$ are observed in an experiment). In this case the sums are expressions of the sum rule, Eq. (2.23),

$$\sum_{N=-11}^{+11} R_N = 1$$

It was explained in Section II.C that Eq. (2.24) can be used to extract properties of the laser out of measured FF cross sections from the directly observable quantities R_N. This procedure was amply discussed by Jung (1980), who also gives full details of the derivation of the following relation:

$$\langle F \rangle = \frac{\hbar^2 \omega^4 cm^2}{4\pi e^2 (\varepsilon \cdot \mathbf{q})^2} \sum_{N=-11}^{+11} N^2 R_N$$

where m and e are the mass and the charge of the electron respectively, and

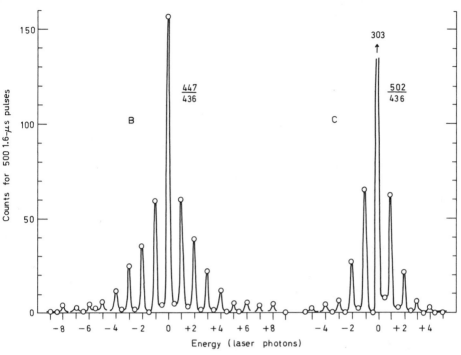

Fig. 6. Multiphoton FF electron energy spectra for e^-–Ar scattering measured in the presence of the flux densities contained in the time slices B and C of the laser pulse shown in Fig. 2. For additional information see Fig. 5.

q is the momentum transfer ($\mathbf{p}_f - \mathbf{p}_i$). This formula allows the calculation of the average flux density $\langle F \rangle$ in absolute units from the directly observable values of R_N. An example of its application is given by Jung (1980) and the same formula was also applied to the latest data. The calculated average flux densities for Figs. 5 and 6 corresponding to pulse time slices A, B, and C in Fig. 2 are as follows:

Channel A: 1.04×10^8 W cm^{-2}

Channel B: 4.5×10^7 W cm^{-2}

Channel C: 9.5×10^6 W cm^{-2}

Other examples have been reported (Weingartshofer *et al.*, 1982). The average of all the calculated values were used to reconstruct the laser pulse shown in Fig. 2, and it is remarkable how well this information fits the measured profile from photographs taken with a fast 400-MHz oscilloscope. This has permitted a direct calibration of the flux density scale in the target region in absolute units given along the ordinate in Fig. 2. The total energy in the pulse has been estimated as 11.2 J which is in very good agreement with the measured laser output of 12 J ($\pm 10\%$) with a joulemeter.

The quantitative results on the sum rule and the average value of the flux density are only indirect tests of the Kroll–Watson formula. They represent the sum or the average of the peak structure of the FF cross section They are global results and do not reflect subtle changes in the peak structure, i.e., asymmetries (absorption versus emission). Take the sum rule for instance, 24 peaks (R_N values) are added up to give a value close to unity, as it should be. The same can be said in the calculation of $\langle F \rangle$, here each R_N value is multiplied with N^2 and 24 products are summed up. Again the agreement is very reasonable. Such tests represent a very important advancement in the instrumentation and measuring technique of FF cross sections.

In this presentation of experimental results we have showed that there has been a continuous upgrading of the instrument since the first measurements of FF transitions. It is still more interesting to realise that great possibilities exist for further improvements, which may very well open the way for the investigation of subtle changes and effects in FF cross sections that have been discussed throughout the text.

D. Projected Experimental Improvements

All the reported results have been obtained with a multimode laser. Although we know that not many modes contribute to the beam [contrary to the description of the chaotic laser field by Zoller (1980)] the laser pulses (e.g., Fig. 2) show strong intensity variations in time and also in space. The detector collects signals from events which have occurred in different space–time points and therefore in quite different field intensities. This presents serious problems that have been discussed in Section II.C. Therefore, the first improvement considered for future measurements will be to remedy this complication as best as possible. A new laser system has been acquired (Lumonics Inc.) consisting of one hybrid laser followed by two amplifiers. Although it should bring a substantial improvement in the measurements, the direct test of the Kroll–Watson formula will still not be possible since this would require a single mode and a very homogenous laser pulse in space and time, ideally square pulses.

We have described a fast digital counter having four digital display channels A, B, C, and D. A new counter has now been constructed with 2000 50-ns time intervals to replace the four-channel counter (A, B, C, and D). This counter will assist in the exact location of the onset of channel A (i.e., time of arrival of the first electrons that have been affected by the intense spike of the laser pulse). It is desirable to monitor the effect of the total flux density of the laser pulse and correlate the shape of the laser pulse with the FF cross sections with discrimination of undesirable laser shots.

The global effects of the theory have been tested by running experiments that have lasted up to 24 hours. This presents problems in stability, alignment, and drift in electron energy. Long experiments are also costly in gas consumption, particularly for the future operation of the three units in the new laser system mentioned above. A research team of the University of Manchester (Hicks *et al.*, 1980) has recently developed a position sensitive multidetector that can accumulate signals from a wide range of electron energies simultaneously. The group has demonstrated an improvement of more than a factor of 100 in sensitivity over the best existing spectrometers. This new innovation is ideally suited for the measurement of FF cross sections. A version of this multidetector should drastically shorten the collection time of data and greatly increase the amount of collected data since it will record many intermediate points between the peaks and valleys that are shown in Figs. 5 and 6. At least two orders of magnitude in the improvement of statistical errors can be expected. The small photon energy of the CO_2 laser is of great advantage making it possible to record the spectrum shown in Fig. 5 in one experimental run with the multidetector since the electron energy distribution occupies an energy range of approximately 2.6 eV (22×0.117).

Pulsed gas beam sources are now commercially available. They have the advantage of producing very high gas densities in the target region while the pressure throughout the vacuum chamber remains low. They have a duration of 150 μs or longer. These pulsed gas beams are of particular interest in our experiments where one component, the laser, is already pulsed.

E. Free—Free Measurements Including Resonances

This survey would be incomplete without presenting the classical results reported by Andrick and Langhans (1977) and Langhans (1978) on the 1-photon resonance structures in the FF cross section of e⁻–Ar scattering. The authors have measured the excitation function for electrons having lost the energy of one photon by stimulated emission in the region around the well-known doublet $^2P_{3/2,1/2}$ resonances in Ar near 11 eV. For each of the temporary negative ion states they find two structures in the excitation function in the presence of the laser field (50-W cw laser focused to about 2.5 mm², i.e., 2×10^3 W cm⁻²). The results are shown in Fig. 7. This is in agreement with theoretical predictions (Section II.C and Krüger and Schulz, 1976). The upper curve is the measured elastic cross section, which is compared with a computed curve. It is *not* a fit, but a calculation from scattering phase shifts using the Fano formalism and with an apparatus

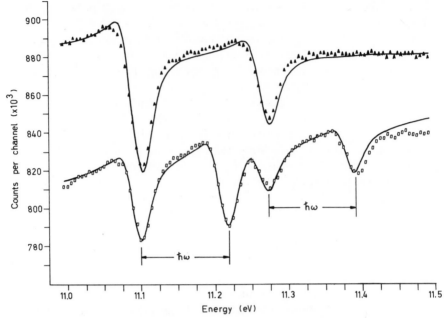

Fig. 7. Experimental comparison of resonance structures in elastic scattering and in one-photon emission. Upper curve: elastic cross section; full curve calculated, full triangles measured. Lower curve: cross section for one-photon emission; full curve calculated, open squares measured. [Reprinted with permission from Langhans (1978). Copyright 1978 The Institute of Physics.]

function folded in. Only a scaling factor and the instrumental width have been adjusted. The same procedure has been applied to the lower curve which corresponds to the FF cross section. The agreement is good and is a further support of the lfa.

It took nearly 100 hours to measure the FF cross section curve. Measurements of this nature have not yet been attempted with intense laser fields.

IV. SUGGESTIONS FOR FUTURE WORK

To our knowledge, no theoretical treatment of multiphoton FF transitions in strong fields has yet been presented which does not use either some type of lfa or a perturbation expansion of the electron–field interaction or a Born series expansion of the electron–target interaction. Therefore not much is known about the magnitude of the error of the lfa in function of such parameters as incoming electron energy, scattering angle, laser frequency, polarization, intensity, etc.

In order to test for errors in the lfa, it would also be useful to develop a numerical treatment of FF transitions which is nonperturbative with respect to the electron–laser and electron–target interactions and which does not use the lfa. This could also give some indications of under which circumstances it is possible to observe a breakdown of the lfa in an experiment and to observe off-shell effects of the electron–target scattering. This would be a big advancement in so far as it would lead to situations where FF experiments can provide information on the electron–target system, which is not available in radiationless electron–target experiments.

The direct experimental test of the Kroll–Watson formula will probably remain impossible in the near future because of the experimental difficulties in the preparation of a homogeneous laser beam. Jung (1981) has suggested a way to circumvent these problems by checking various experiments, done with the same laser beam, against each other. What are needed to realize this idea are experiments carried out with extremely good statistics in the electron counts and therefore also a high total number of counts. This cross check of experimental data against each other could be tried with measurements of the kind done by Weingartshofer *et al.*, except performed with improved statistics. This could provide an indirect verification of Eq. (2.1). Another test of the theory could be to measure resonance profiles in FF transitions and compare them with the corresponding resonance profiles in radiationless scattering. The results of this comparison could be checked for compatibility with the predictions of the resonant version of the lfa. It would be sufficient to repeat the measurements of Andrick and Langhans for various values of the scattering angle and perhaps also for various laser intensities. The measurements could also be done with different types of lasers with different photon energies in order to see a breakdown of the lfa for sufficiently high photon energies.

Finally, it would be definitely desirable to perform more complicated types of FF transitions involving target excitation, target ionization, electron spin-flip, etc., in the presence of an external field.

ACKNOWLEDGMENTS

One of the authors (A.W.) wishes to express appreciation to Dr. J. Sabbagh and Mr. J. K. Holmes for assistance in the experimental aspects of this review.

We gratefully acknowledge the continuous support of the National Sciences and Engineering Research Council of Canada, and express our thanks to the St. Francis Xavier Council for Research for a generous grant.

REFERENCES

Andrick, D., and Langhans, L. (1977). *Abstr. Pap. Int. Conf. Phys. Electron. At. Collisions, 5th*, pp. 350–351.
Banerji, J., and Mittleman, M. H. (1981). *J. Phys. B* **14**, 3717–3725.
Banerji, J., and Mittleman, M. H. (1982). *Phys. Rev. A* **26**, 3706–3708.

Beaulieu, J. A. (1970). *Appl. Phys. Lett.* **16**, 504–505.
Bunkin, F. V., and Fedorov, M. V. (1966). *Sov. Phys. JETP (Engl. Transl.)* **22**, 844–847.
Cavaliere, P., Ferrante, G., and Leone, C. (1980). *J. Phys. B* **13**, 4495–4507.
Chandrasekhar, S., and Breen, F. H. (1946). *Astrophys. J.* **104**, 430–445.
Conneely, M. J., and Geltman, S. (1981). *J. Phys. B* **14**, 4847–4864.
Coulter, P. W., and Ritchie, B. (1981). *Phys. Rev. A* **24**, 3051–3060.
Daniele, R., Ferrante, G., and Bivona, S. (1981). *J. Phys. B* **14**, L213–L218.
Dyachkov, L. G., Kobzev, G. A., and Norman, G. E. (1975). *Sov. Phys. JETP (Engl. Transl.)* **38**, 697–700.
Feshbach, H., and Yennie, D. R. (1962). *Nucl. Phys.* **37**, 150–171.
Gallagher, J. W. (1977). "Bibliography of Free–Free Transitions in Atoms and Molecules." JILA Information Center, Report No. 16, Boulder, Colorado.
Gradshteyn, I. S., and Ryzhik, I. M. (1965). "Table of Integrals, Series and Products." Academic Press, New York.
Gavrila, M. (1977). *Invited Pap. Prog. Rep. Int. Conf. Phys. Electron. At. Collisions, 10th,* pp. 165–184.
Gavrila, M., and Van der Wiel, M. (1978). *Comments At. Mol. Phys.* **8**, 1.
Geltman, S. (1972). *J. Quant. Spectrosc. Radiat. Transfer* **13**, 601–613.
Geltman, S. (1977). *J. Res. Nat. Bur. Stand.* **82**, 173–179.
George, T. F. (1982). *J. Phys. Chem.* **86**, 10–20.
Gradshteyn, I. S., and Ryzhik, I. M. (1965). "Table of Integrals, Series, and Products." Academic Press, New York.
Heller, L. (1968). *Phys. Rev.* **174**, 1580–1587.
Hicks, P. J., Daviel, S., Wallbank, B., and Comer, J. (1980). *J. Phys. E* **13**, 713–715.
Jauch, J. M., and Rohrlich, F. (1976). "The Theory of Photons and Electrons." Springer-Verlag, Berlin and New York.
Jung, C. (1979). *Phys. Rev. A* **20**, 1585–1589.
Jung, C. (1980). *Phys. Rev. A* **21**, 408–411.
Jung, C. (1981). *Phys. Rev. A* **24**, 360–369.
Jung, C., and Krüger, H. (1978). *Z. Phys. A* **287**, 7–13.
Jung, C., and Krüger, H. (1979). *Can. J. Phys.* **57**, 1792–1799.
Jung, C., and Taylor, H. S. (1981). *Phys. Rev. A* **23**, 1115–1126.
Keen, B. E. (1974). *Conf. Ser. Inst. Phys.* No. 20.
Key, M. H., and Hutcheon, R. J. (1980). *Adv. At. Mol. Phys.* **16**, 201–280. (In particular, see page 249.)
Kroll, N. M., and Watson, K. M. (1973). *Phys. Rev. A* **8**, 804–809.
Krüger, H., and Jung, C. (1978). *Phys. Rev. A.* **17**, 1706–1712.
Krüger, H., and Schulz, M. (1976). *J. Phys. B* **9**, 1899–1910.
Langhans, L. (1978). *J. Phys. B* **13**, 2361–2366.
Langendam, P. J. K., and Van der Wiel, M. J. (1978). *J. Phys. B* **11**, 3603–3613.
Low, F. E. (1958). *Phys. Rev.* **110**, 974–977.
Mittleman, M. H. (1979a). *Phys. Rev. A* **19**, 134–138.
Mittleman, M. H. (1979b). *Phys. Rev. A* **20**, 1965–1971.
Mittleman, M. H. (1980). *Phys. Rev. A* **21**, 79–84.
Mittleman, M. H. (1982). *Comments At. Mol. Phys.* **11**, 91–100.
Morgan, C. G. (1975). *Rep. Prog. Phys.* **38**, 621–665.
Nordsieck, A. (1937). *Phys. Rev.* **52**, 59–62.
Rosenberg, A., Felsteiner, J., Ben-Arych, Y., and Politch, J. (1980). *Phys. Rev. Lett.* **22**, 1787–1790.
Rosenberg, L. (1979a). *Phys. Rev. A* **20**, 275–280.
Rosenberg, L. (1979b). *Phys. Rev. A* **20**, 1352–1358.

Rosenberg, L. (1979c). *Phys. Rev. A* **20**, 457–464.
Rosenberg, L. (1981). *Phys. Rev. A* **23**, 2283–2292.
Schlessinger, L., and Wright, J. (1979). *Phys. Rev. A* **20**, 1934–1944.
Shakeshaft, R., and Robinson, E. J. (1982). *Phys. Rev. A* **25**, 1977–1985.
Sommerfeld, A., and Maue, A. W. (1935). *Ann. Phys.* **23**, 589–596.
Weingartshofer, A., and Jung, C. (1979). *Phys. Can.* **35**, 119–124.
Weingartshofer, A., Willmann, K., Clarke, E. M. (1974). *J. Phys. B* **7**, 79–90.
Weingartshofer, A., Holmes, J. K. Caudle, G., Clarke, E. M., and Krüger, H. (1977). *Phys. Rev. Lett.* **39**, 269–270.
Weingartshofer, A., Clarke, E. M., Holmes, J. K., and Jung, C. (1979). *Phys. Rev. A* **19**, 2371–2376.
Weingartshofer, A., Holmes, J. K., and Sabbagh, J. (1981). *Laser Spectros., Proc. Int. Conf., 5th*, pp. 247–250.
Weingartshofer, A., Holmes, J. K., Sabbagh, J., and Chin, S. L. (1983). *J. Phys. B* **16**, 1805–1817.
Zoller, P. (1980). *J. Phys. B* **13**, L249–L252.

8

Multiphoton Autoionization

P. LAMBROPOULOS

Department of Physics
University of Southern California, University Park
Los Angeles, California
and
University of Crete
Iraklion, Crete, Greece

P. ZOLLER

Institut für Theoretische Physik
Universität Innsbruck
Innrain, Innsbruck, Austria

I. INTRODUCTION

Considerable knowledge on multiphoton ionization of atoms has accumulated by now as attested by the preceding chapters of this volume.
Most of the knowledge and certainly the best understood part, concerns processes in which one electron is ejected in a transition which leaves the core

more or less unaffected. The alkali atoms and, of course, hydrogen are the examples that best satisfy this condition, and it is from investigations on such atoms that the evidence has been collected. Even in the one-electron picture of these atoms, the calculational task involved in a multiphoton generalized cross section—especially of order higher than 3—is formidable. As a result, there are no comparisons between theory and experiment for order higher than 5, and those for 4 and 5 are extremely few and with one exception rather unsettled. Experimentally, on the other hand, high-order (up to 22) multiphoton ionization of rare gases has been found to behave as would be expected on the basis of a one-electron model, at least in so far as the dependence on laser power is concerned. Whether the model is adequate for all aspects of these processes is a question to which we return below. The above difficulties notwithstanding, the accumulated experience has led to the formation of an intuition that enables one to make a zeroth-order guess about the outcome of a multiphoton ionization experiment on any atom, or so it seemed until recently.

New experimental information [1–5] that began appearing in 1979 seems to suggest that the above picture may be part of a larger landscape. As discussed in previous chapters, it has been found that the photoelectron under certain conditions can keep absorbing photons in the continuum [4]. It happens that this has so far been observed only in rare gases. On the other hand, it has also been observed that in multiphoton ionization of rare gases multiply ionized specied appear [5]. The simplest composite picture of these two types of observation would be the sequential ionization of two electrons. The first is ejected, absorbing photons in the continuum as well, while the second begins, so to speak, its transition after the singly ionized ion has been created. But if one of the electrons is absorbing photons in the continuum, it is essentially meaningless to talk about the first ion having been created first. We are thus quickly led to the admission that it may not be always possible to think of the two electrons as undergoing transitions independently. Obviously the picture becomes even more complex when we attempt to account for the appearance of triple and quadruple ionized species which is by now an experimental fact [5]. Given that the rare gases have closed shells, should one be surprised if under certain circumstances more than one electron is involved in a transition under a strong field?

Although the theoretical understanding of these phenomena in the rare gases is at a primitive stage, different but related types of experiments are more amenable to some quantitative interpretation. They have to do with multiphoton ionization of alkaline earths [1–3], and certain aspects of the existing data have been discussed in the preceding chapters. The alkaline earths have two-electron valence shells. They also have a plethora of autoionizing states above the first ionization threshold. These can be thought of as two-electron excited configurations embedded in the continuum. Such autoionizing states are known either from vacuum ultraviolet (vuv) photo-absorption or electron scattering. In recent years, multiphoton absorption studies [6, 7] have also been employed as a tool for the exploration of that energy range of the alkaline earths. The additional information provided by these studies consists of the higher angular momentum states that could not be reached via single-photon absorption. It is thus expected that in multi-photon ionization of these atoms, such autoionizing states will be reached and will probably play an important role in the total ionization. Moreover, since two-electron excited states are involved, it is conceivable that a transition via more than one autoionizing state can lead to the ejection of both electrons and hence to doubly ionized species. Doubly ionized alkaline earths have been observed, as we know from the previous chapters. They have been observed with surprisingly low laser intensities if one considers the rather high-order process involved in ionizing the ground state of the singly ionized ion.

Looking now at the situation in rare gases and alkaline earths compar-atively, we see that in both, the continuum participates in a way much more complex than in the previous generation of multiphoton experiments. Since the autoionizing structure in alkaline earths is expected to lie relatively close to threshold, it is not too surprising that two-electron excitation and ion-ization has been seen. In rare gases the known autoionization structure is somewhat different and not as dense. Of course, what such structure would look like for higher energy of excitation and high angular momenta is totally unknown. There surely must be two-electron excitations participating in those experiments as well, but that is one of the important questions in the field. On the other hand, the absorption of photons in the continuum (often called free–free absorption) by the single electron that has been observed [4] in Xe has not been found either in alkaline earth or alkali atoms. That is not quite true since the additional absorption of photons above the first thresh-old has been seen in alkaline earths, except that there it can be correlated with the presence of doubly excited states. But it has been seen in the alkalis where the continuum has no low-lying structure. It could be simply a matter of experimental conditions. It could, however, have to do with the different potential that the single electron just above threshold sees in the two cases. In an alkali it sees a closed shell, while in a rare gas it sees a shell with a va-cancy containing an unpaired electron. It is known that autoionizing states due to so-called angular momentum recoupling (change of angular momen-tum of the core) do exist, although not in the appropriate energy range for the multiphoton experiments. It is therefore an open question whether even the additional absorptions in the continuum of the rare gases is simply a phenomenon of one electron. Core participation of some form (other than just providing a long-range attraction) may well be involved there as well.

The central theme that appears to run through the above discussion of the new observations is core participation in the multiphoton process. The core is no longer a spectator providing a long-range field, but it undergoes excitations that either affect the behavior of the ejected electron or lead to the ejection of an additional electron, and so on. Understanding the behavior of two-electron excited states is then an important and necessary first step towards the unraveling of the variety of new mysteries that multiphoton ionization presents in its new phase. Such states can participate as final autoionizing states as intermediate states or combinations thereof. Especially when absorptions in the continuum are involved, these states could enter as either real or virtual (nonresonant) intermediate states in the overall transition. Given that the observations are made under rather strong laser fields, the problem is not simply the construction of two-electron states of the bare atom but also their possible modification by the field. After all, two-electron states involve the electron–electron repulsion, which is the fundamental agent of configuration interaction. It is natural and in keeping with existing experience to ponder the conditions under which the effect of the field becomes comparable to that of configuration interaction. As we will see in subsequent pages of this chapter, it takes not too large fields, and certainly is within the range of current experiments, for this to occur. Then the process can not be thought of as taking place via the states of the bare atom (often referred to as the physical states) but via the states that may look significantly different. The task therefore is not only the correct identification and construction of the two-electron excited states, but also their construction, with the field taken into account on an equal footing.

The problem is relatively new and considerably more complex than that of multiphoton ionization of a single-electron atom. In fact, it contains all the difficulties of the latter, involves all of the ingredients of the atomic physics of two-electron systems, and, in addition, requires the inclusion of the field not as a small perturbation or a weak transition-causing probe. A beginning has been made and the discussion that follows in the next few sections attempts to set up a framework for addressing the problem. The existing work, as of the writing of this chapter, is mostly formal and some of it too simplified to be of direct relevance to the realistic physical situation. It is not even clear at this point whether the approaches presently in the literature are the best for the purpose. But the essential aspects of the thinking and formalism contained in the following pages are likely to survive as a way of modeling the physical problem and to continue being useful at least in particular experimental contexts.

II. FORMAL THEORY

A. Weak-Field Theory

The standard formulation [8] of a transition involving an autoionizing state in photoabsorption is cast in terms of a transition probability per unit time. The relevant matrix element involves an initial state $|g\rangle$, the dipole operator

$$D \equiv -e(\mathbf{r} \cdot \lambda)\mathscr{E}(t) \equiv (\boldsymbol{\mu} \cdot \lambda)\mathscr{E}(t) \qquad (2.1)$$

where λ is the polarization vector and $\mathscr{E}(t)$ is the amplitude of the field, and $\Psi_{\tilde{E}}$ is the autoionizing state with $\tilde{E}$ being its energy which lies in the continuum. If ω is the frequency of the absorbed photon, $\tilde{E}$ is equal to $E_g + \hbar\omega$ and the quantity $|\langle\Psi_{\tilde{E}}|D|g\rangle|^2$ as a function of ω gives the line shape of the absorption, which can be monitored by direct measurement of the attenuation of the incoming radiation, observation of the ejected electron, or combinations and variations thereof. The atomic physics content is in the construction of $\Psi_{\tilde{E}}$ and there is a vast literature [8–14] on the subject. The fundamental feature of the process that must be accounted for by the theory is that $|\langle\Psi_{\tilde{E}}|D|g\rangle|^2$ as a function of ω exhibits a resonance structure whose asymmetry or lack of it reflects the details of the atomic structure at that energy range. The position of the resonance, its width, and the dimensionless asymmetry parameter q are the principal parameters most often appearing in the description of this process. If the full width is denoted by Γ and the center of the resonance by $\bar{E}_a$, the line shape can be expressed as

$$R^2 = (q + \varepsilon)^2/(1 + \varepsilon^2) \qquad (2.2a)$$

where

$$\varepsilon = (\tilde{E} - \bar{E}_a)/\tfrac{1}{2}\Gamma \qquad (2.2b)$$

The above parameters are obtained if $\Psi_{\tilde{E}}$ is known.

The task of obtaining a good $\Psi_{\tilde{E}}$ is, of course, the hardest part of the atomic structure calculation for this problem, and there exists may approaches in the literature [8–14]. In all approaches the essential feature embodied in $\Psi_{\tilde{E}}$ is that it contains a continuum as well as a discrete part in an appropriate superposition. The representation and calculation of these wave functions is not unique because it involves approximations that seldom can be evaluated with high accuracy. As a result, there always exists a significant dose of judgment as to which model is best for a specific problem. Let us, for the time being, assume that these wave functions are known and let $|a\rangle$ with energy E_a and $|c\rangle$ with energy E_c be the discrete and continuous parts, respectively, in some zeroth-order approximation. There is an interaction V that couples $|a\rangle$ and $|c\rangle$ through the matrix element $\langle c|V|a\rangle \equiv V_{ca}$. It is the interaction that causes autoionization, or at least part of it. The typical way [8] of obtaining $\Psi_{\tilde{E}}$ is to diagonalize the Hamiltonian $H^A + V$ (where $|a\rangle$, $|c\rangle$, and $|g\rangle$ are assumed eigenstates of H^A) in the space spanned by $|a\rangle$ and $|c\rangle$. The resulting expression for $\Psi_{\tilde{E}}$ is

$$|\Psi_{\tilde{E}}\rangle = \frac{\sin \Delta}{\pi V_{\tilde{E}}}|\Phi_{\tilde{E}}\rangle - \cos \Delta |c\rangle \qquad (2.3a)$$

where

$$|\Phi_{\tilde{E}}\rangle = |a\rangle + \mathbb{P}\int dE_c \frac{V_{ca}}{\tilde{E} - E_c}|c\rangle \qquad (2.3b)$$

The matrix element V is of course a function of E_c, and $V_{\tilde{E}}$ stands for $V_{ca}(E_c = \tilde{E})$, i.e., V_{ca} is evaluated at $\tilde{E}$ and the integral in Eq. (2.3b) is to be understood as a principal value indicated by $\mathbb{P}$. The quantity Δ is defined by

$$\Delta = -\arctan \pi |V_{\tilde{E}}|^2/(\tilde{E} - \bar{E}_a) \qquad (2.4)$$

where $\bar{E}_a$ is the energy E_a modified by a shift due to the interaction V. The width Γ is also expressed in terms of $V_{\tilde{E}}$ as

$$\tfrac{1}{2}\Gamma = \pi |V_{\tilde{E}}|^2 \qquad (2.5)$$

All of the states employed in such formalisms are to be understood as two-electron states properly antisymmetrized. Conceptually the simplest way of thinking about $|a\rangle$ and $|c\rangle$ would be as two-electron states without the electrostatic electron–electron interaction e^2/r_{12}. Thus V represents just this interaction and the matrix elements of V are matrix elements of e^2/r_{12}. This is the underlying picture in Fano's treatment where the eigenstates of H^A are also orthogonal to each other. Computationally, for atoms other than He, this scheme is impractical. One must think in terms of states $|a\rangle$, which are (approximate) eigenstates of a portion of $H^A + V$ that contains part of e^2/r_{12}. In addition, $|a\rangle$ and $|c\rangle$ may not necessarily be orthogonal. In that case, autoionization is owing to the remaining part of e^2/r_{12} and to the overlap $\langle a|c\rangle$. Of course, all these different ways express, in schemes differing calculationally, the underlying physical effect, which is the ejection of one of the two excited electrons because of electrostatic repulsion. The differences between such models are by no means trivial, and they may in fact introduce conceptual nuances in the way one pictures the process. Since the focus of our attention here is the effect of the field on the processes, we will not elaborate on these aspects. We simply adopt the model represented by Eqs. (2.3)–(2.5) because it contains the features that are essential to our purpose. For example, nonorthogonality of $|a\rangle$ and $|c\rangle$ does modify the equations but not in an essential way.

To complete this summary of the standard formalism, let us note that q is given by

$$q(\tilde{E}) \equiv \frac{\langle \Phi_{\tilde{E}} | D | g \rangle}{\pi V_{\tilde{E}} \langle c | D | g \rangle_{\tilde{E}}} \equiv \frac{\tilde{D}_{\tilde{E}g}}{\pi V_{\tilde{E}} D_{cg}(\tilde{E})} \tag{2.6}$$

When $|\Psi_{\tilde{E}}\rangle$ is coupled to $|g\rangle$ via a single-photon absorption, inspection of Eq. (2.3) shows that the transition has two paths: one directly to $|c\rangle$ and the other to $|\Phi_{\tilde{E}}\rangle$. The parameter q is a measure of the interference between the two and determines the degree of asymmetry. Large q implies weak transition to $|c\rangle$ and hence symmetric line shape, while small q implies asymmetric line shape.

B. Strong-Field Theory: Formulation

If we are to consider situations in which the radiation field is sufficiently strong for the quantity $\tilde{D}_{\tilde{E}g}$ to become comparable to Γ, a different treatment is necessary. The field is no longer a weak probe and the process (ion yield, line shape, etc.) may not be describable in terms of a simple transition probability for unit time. A formulation treating the coupling of the field with the dominant atomic states in a nonperturbative way is then necessary. By dominant atomic states we mean those states that make the major contribution to the problem owing to the particular photon frequency, order of multiphoton process, etc., very much like the way V was handled in Section II.A. The Hamiltonian $H^0 + V$ was prediagonalized in a restricted space spanned by the states strongly coupled by V, namely $|a\rangle$ and $|c\rangle$. If we stay with the same problem for the moment (one autoionizing resonance reached by a single-photon transition from an initial state), we recognize immediately that the space of strongly coupled states must include $|g\rangle$ because strong field entails strong coupling of $|g\rangle$ to $|a\rangle$ and $|c\rangle$. This can be accomplished by treating the field either classically or quantum mechanically, and both approaches can be found in Ref. [15]. It is slightly more convenient for our discussion here to employ the fully quantized version.

The total Hamiltonian now is

$$H = H^A + H^R + V + D \equiv H^0 + V + D \tag{2.7}$$

where H^A again denotes (zeroth-order) atomic part with eigenstates $|g\rangle$, $|a\rangle$, and the continuum $\{|c\rangle\}$, and the radiation part H^R with eigenstates the usual photon-number states $|n\rangle$. Only one mode of the field of frequency ω is assumed occupied. The interactions V and D retain their previous meaning except that D is now written in terms of creation and annihilation operators as

$$D = -ei(2\pi\hbar\omega)^{1/2}[a(t)(\mathbf{r} \cdot \boldsymbol{\lambda}) - a^\dagger(t)(\mathbf{r} \cdot \boldsymbol{\lambda}^*)] \tag{2.8}$$

The initial state of the system is $|g;n\rangle$ to be also denoted by $|g'\rangle$, with energy $E' = E_g + n\hbar\omega$. The states connected to it via the absorption of a photon are $|a;n - 1\rangle \equiv |a'\rangle$ with energy $E'_a + (n - 1)\hbar\omega$ and $|c;n - 1\rangle \equiv |c'\rangle$, with energy $E'_c = E_c + (n - 1)\hbar\omega$. By considering only the above states and ignoring those with $n + 1$ photons, we are simply adopting the standard rotating wave approximation.

The evolution of the wave function is given by

$$\Psi(t) = \exp\left[-(i/\hbar)Ht\right]\Psi(0) \equiv U(t)|g;n\rangle \tag{2.9}$$

thus defining the time-evolution operator $U(t)$, which is expressed in terms of the resolvent operator $G(z)$ as

$$G(z) \equiv \frac{1}{z - H} = \frac{1}{z - H^0 - V - D} \tag{2.10}$$

through the inversion integral

$$U(t) = -\frac{1}{2\pi i}\int_{-\infty}^{+\infty} dx\, e^{-ixt}G^+(x) \tag{2.11}$$

where as usual $G^+(x) = \lim_{\eta \to +0} G(x + i\eta)$. Taking into account the states of interest, $\Psi(t)$ is also written as

$$\Psi = U_{g'g'}|g'\rangle + U_{a'g'}|a'\rangle + \int dE'_c\, U_{c'g'}|c'\rangle \tag{2.12}$$

where the time dependence is in the matrix elements of U, which can in turn be obtained from the corresponding matrix elements of G. For the matrix elements of G we consider Eq. (2.10) written as $(z - H^0)G + VG + DG = 1$ and we simply take the $g'g'$, $a'g'$, and $c'g'$ matrix elements of both sides, noting in the process that the only nonvanishing matrix elements of V and D are $V_{a'c'}$, $D_{a'g'}$, and $D_{c'g'}$. The resulting equations are

$$(z - E'_g)G_{g'g'} - D_{g'a'}G_{a'g'} - \int dE'_c\, D_{g'c'}G_{c'g'} = 1 \tag{2.13a}$$

$$-D_{a'g'}G_{g'g'} + (z - E'_a)G_{a'g'} - \int dE'_c\, V_{a'c'}G_{c'g'} = 0 \tag{2.13b}$$

$$-D_{c'g'}G_{g'g'} - V_{c'a'}G_{a'g'} + (z - E'_c)G_{c'g'} = 0 \tag{2.13c}$$

Since only energy differences matter in the transition, we can take, without any further approximation, $E'_g = E_g + \hbar\omega$, $E'_a = E_a$, and $E'_c = E_c$. Moreover, we note that $V_{a'c'} = V_{ac}$, $D_{a'g'} = \mathscr{E}\mu_{ag}$, and $D_{c'g'} = \mathscr{E}\mu_{cg}$, where μ is to be understood as the projection of $\boldsymbol{\mu}$ on the polarization vector $\boldsymbol{\lambda}$ (see Eq. (2.1)].

We can now simplify the notation by simply using G_g, G_a, and G_c for the matrix elements of G because they basically refer to atomic states. The radiation states have given rise to (the strength of the field in) the matrix elements of D.

From Eq. (2.13c), we can write

$$G_c = \frac{1}{z - E_c} (D_{cg}G_g + V_{ca}G_a) \tag{2.14}$$

which substituted into the other equations leads to

$$\left(z - E_g - \hbar\omega - \int dE_c \frac{|D_{cg}|^2}{z - E_c} \right) G_g - \left(D_{ga} + \int dE_c \frac{D_{gc}V_{ca}}{z - E_c} \right) G_a = 1 \tag{2.15a}$$

$$-\left(D_{ag} + \int dE_c \frac{V_{ac}D_{cg}}{z - E_c} \right) G_g + \left(z - E_a - \int dE_c \frac{|V_{ac}|^2}{z - E_c} \right) G_a = 0 \tag{2.15b}$$

From these equations, we obtain

$$G_g = 1/\Lambda(z) \tag{2.16a}$$

and

$$G_a = \frac{D_{ag} + \int dE_c [V_{ac}D_{cg}/(z - E_c)}{\{z - E_a - \int dE_c [|V_{ac}|^2/(z - E_c)]\}\Lambda(z)} \tag{2.16b}$$

where Λ is defined by

$$\Lambda(z) \equiv z - E_g - \hbar\omega - \int dE_c \frac{|D_{cg}|^2}{z - E_c}$$

$$- \frac{\left(D_{ga} + \int dE_c \frac{D_{gc}V_{ca}}{z - E_c} \right)\left(D_{ag} + \int dE_c \frac{V_{ac}D_{cg}}{z - E_c} \right)}{z - E_a - \int dE_c \frac{|V_{ac}|^2}{z - E_c}} \tag{2.17}$$

Under the assumption that the matrix elements under the integrals are slowly varying functions of E_c over the range of the resonance—an assumption also made in the standard theory—z can be replaced by $E_g + \hbar\omega + i\eta$ under the integrals. Using a well-known identity, we then have

$$\frac{1}{z - E_c} \simeq \lim_{\eta \to +0} \frac{1}{E_g + \hbar\omega - E_c + i\eta}$$

$$= \mathbb{P} \int \frac{dE_c}{E_g + \hbar\omega - E_c} - i\pi\delta(E_g + \hbar\omega - E_c) \tag{2.18}$$

where $\mathbb{P}$ denotes the principal value. This replacement is equivalent to what, in some papers, is referred to as the pole approximation. Substitution into the integrals leads to shifts and widths according to the following equations:

$$\int dE_c \frac{|D_{cg}|^2}{z - E_c} \cong \mathbb{P} \int dE_c \frac{|D_{cg}|^2}{E_g + \hbar\omega - E_c} - i\pi |D_{cg}(E_g + \hbar\omega)|^2 \equiv S_g - \frac{i}{2}\gamma_g \quad (2.19)$$

$$\int dE_c \frac{|V_{ca}|^2}{z - E_c} \cong \mathbb{P} \int dE_c \frac{|V_{ca}|^2}{E_g + \hbar\omega - E_c} - i\pi |V_{ca}(E_g + \hbar\omega)|^2 \equiv F_a - \frac{i}{2}\Gamma_a \quad (2.20)$$

We also obtain the relation

$$D_{ag} + \int dE_c \frac{V_{ac}D_{cg}}{z - E_g} \cong D_{ag} + \mathbb{P} \int dE_c \frac{V_{ac}D_{cg}}{E_g + \hbar\omega - E_c} - i\pi(V_{ac}D_{cg})_{E_g + \hbar\omega}$$

$$\equiv \tilde{D}_{E_g}(1 - i/q) \quad (2.21)$$

where $\tilde{D}_{\tilde{E}_g}$ and q are the same parameters we defined in Section II.A. The quantities S_g and γ_g are, respectively, the shift and the ionization width of $|g\rangle$ due to its coupling to the continuum by the radiation. Strictly speaking S_g should also include a summation over bound states connected to $|g\rangle$ via a nonvanishing dipole matrix element. They are not included here because they are absent from our initial model. Such shifts do not appear in the standard theory because they are negligible due to the weakness of the field. Here they must be included. The shift F_a and width Γ_a are owing to the interaction that causes autoionization and are exactly the quantities of the standard theory as defined, for example, by Fano [5]. In that case Γ_a turns out to be the width of the autoionizing resonance; assuming $|a\rangle$ is orthogonal to $|c\rangle$. For strong field, Γ_a is not necessarily the width of the observed resonance as will be shown. It is one of the parameters that affect the width together with γ_g and $\tilde{D}_{\tilde{E}_g}$. We shall nevertheless continue referring to it as the autoionization width.

Solving Eqs. (2.15) we obtain G_g, G_a, and G_c from Eq. (2.14), from which we calculate $U_g(t)$, $U_a(t)$, and $U_c(t)$. For weak field the usual perturbation theory result is equivalent to calculating the transition probability per unit time as $\lim_{t \to \infty} (1/t)|U_g(t)|^2$. This quantity is not necessarily meaningful for strong field because $|U_g(t)|^2$ may undergo oscillations. The more general and appropriate quantity to calculate then is the total ionization probability $P(t)$ given by

$$P(t) = 1 - |U_g(T)|^2 - U_a(T)e^{-\Gamma_a(t - T)} \quad (2.22)$$

where T is the interaction time (usually the laser duration or else the time spent by the atom in the interaction region, whichever is smaller) and t is the time over which ions (or electrons) are collected. In practice $\Gamma_a(t - T) \gg 1$

and we can take

$$P(T) = 1 - |U_g(T)|^2 \qquad (2.23)$$

as the observed quantity. For further discussion of the difference between (2.22) and (2.23), and the possibility of actually observing the difference with short laser pulses, see Ref. [15]. An expression equivalent to that of Eq. (2.23) for $P(T)$ is

$$P(T) = \int dE_c \, |U_c(T)|^2 \qquad (2.24)$$

which results from probability conservation. We shall then assume hereafter that T is the laser pulse duration. It bears emphasizing that in most experiments with pulsed strong lasers, the time is an important consideration that can significantly affect line shapes and other aspects of the observation; a point to which we return later. Expressions (2.22) or (2.23) will of course reduce to the usual form $1 - e^{-WT}$, where W is the transition probability per unit time when the appropriate conditions are satisfied.

C. Strong-Field Theory: Solutions

Before discussing the solutions of Eqs. (2.15), it will prove useful to rewrite them in a more compact form using the quantities defined by Eqs. (2.19)−(2.21). First, we define a new z translated by $E_g + S_g + \hbar\omega$ and denote it by

$$x = z - (E_g + S_g + \hbar\omega) \qquad (2.25)$$

Second, whenever an energy is modified by a shift, we incorporate it in the energy, thus defining

$$\bar{E}_a \equiv E_a + F_a \quad \text{and} \quad \bar{E}_g \equiv E_g + S_g \qquad (2.26)$$

Finally, we introduce the detuning from resonance defined by

$$\delta = \hbar\omega - (\bar{E}_a - \bar{E}_g) \qquad (2.27)$$

The equations for G_g and G_a can now be written as

$$[x + (i/2)\gamma_g]G_g - \tilde{\Omega}[1 - (i/q)]G_a = 1 \qquad (2.28a)$$

$$-\tilde{\Omega}[1 - (i/q)]G_g + [x + \delta + (i/2)\Gamma_a]G_a = 0 \qquad (2.28b)$$

which show that the essential parameter is the detuning and not the absolute positions of E_g and E_a. The shift F_a is model-dependent since it refers to state $|a\rangle$, which has an element of arbitrariness in its choice. The quantity that can be related to observation is $E_a + F_a$, which should not depend on the model. Thus, in a sense, F_a is a computational artifact. The shift S_g, on the other hand, is real, intensity-dependent (proportional to the laser intensity), and affects the observation. It shifts the position of the observed resonance as the intensity changes. As a result, δ as defined above is also intensity-dependent because it incorporates S_g and it could be called dynamic detuning, a term often used in multiphoton processes. Obviously the equations can be cast in terms of a static detuning δ_0 by simply replacing δ by $\delta_0 + S_g$. The symbol $\tilde{\Omega}$ is the same as $\tilde{D}_{\bar{E}_g}$, and it is useful to recall the relations

$$\tilde{\Omega} \equiv \tilde{D}_{\bar{E}_g} = \mathscr{E}\tilde{\mu}_{\bar{E}_g} = \mathscr{E}\langle\Phi_{\bar{E}}|\mu|g\rangle \tag{2.29}$$

to remind us that $\tilde{\Omega}$ is proportional to the electric field amplitude.

Because of the transformation given by Eq. (2.25), the inversion integrals of $G_g(x)$ and $G_a(x)$ will give the quantities

$$u_g(t) \equiv U_g(t)\exp[(i/\hbar)(\bar{E}_g + \hbar\omega)t] \quad \text{and} \quad u_a(t) \equiv U_a(t)\exp[(i/\hbar)\bar{E}_a t] \tag{2.30}$$

which simply differ from U_g and U_a by phase factors not affecting the populations. From Eqs. (2.28), we now have

$$u_g(t) = \frac{1}{2\pi i}\int_{-\infty}^{+\infty}dx\,\frac{x + \delta + (i/2)\Gamma_a}{\Lambda(x)}e^{-ixt} \tag{2.31a}$$

$$u_a(t) = \frac{1}{2\pi i}\int_{-\infty}^{+\infty}dx\,\frac{\tilde{\Omega}(1 - (i/q))}{\Lambda(x)}e^{-ixt} \tag{2.31b}$$

where, in terms of the new variable x, Λ is given by

$$\Lambda(x) = [x + (i/2)\gamma_g][x + \delta + (i/2)\Gamma_a] - \tilde{\Omega}^2[1 - (i/q)]^2. \tag{2.32}$$

If we denote the roots of $\Lambda(x) = 0$ by $x_\pm$, then

$$x_\pm = -\tfrac{1}{2}[\delta + (i/2)(\gamma_g + \Gamma_a)] \pm \tfrac{1}{2}\{[\delta - (i/2)(\gamma_g - \Gamma_a)]^2 + 4\tilde{\Omega}^2[1 - (i/q)]^2\}^{1/2} \tag{2.33}$$

and the final expression for $u_g(t)$ is

$$u_g(t) = \frac{1}{x_+ - x_-}\{[x_+ + \delta + (i/2)\Gamma_a]e^{-ix_+t} - [x_- + \delta + (i/2)\Gamma_a]e^{-ix_-t}\} \tag{2.34}$$

Substitution of $x_\pm$ above leads to rather complicated expressions, which are exact but of little inspectional usefulness, except for some special cases that we explore below.

For weak field, the standard theory of Section II.A should be obtained. Weak field means $|\tilde{\Omega}| \ll \Gamma_a$, which also implies $|\tilde{\Omega}[1 - (i/q)]| \ll |\delta -$

$(i/2)(\gamma_g + \Gamma_a)|$, and therefore the square root in Eq. (2.33) can be expanded and approximated by

$$1 + \frac{1}{2} \frac{4\alpha^2}{[\delta - (i/2)(\gamma_g - \Gamma_a)]^2}$$

where $\beta^2 \equiv 4\tilde{\Omega}^2/q^2\Gamma_a^2$ and $\alpha^2 \equiv \tilde{\Omega}^2(1 - i/q)^2$. The roots then, to lowest order, reduce to

$$x_+ \cong \frac{\alpha^2}{\delta + (i/2)\Gamma_a} - \frac{i}{2}\beta^2\Gamma_a \qquad (2.35a)$$

and

$$x_- \cong -\left(\delta + \frac{i}{2}\Gamma_a\right) - \frac{\alpha^2}{\delta + (i/2)\Gamma_a} \qquad (2.35b)$$

Because of the smallness of α, $|x_-| \gg |x_+|$ and only the small root x_+ will contribute to the decay rate of $U_g(t)$ (the rate with which $|g\rangle$ ionizes). The rate is $-2\,\mathrm{Im}\,x_+$, which is nothing else but $[(d/dt)|U_g(t)|^2]_{t=0}$. After some straightforward algebra and introducing $\varepsilon \equiv \delta/\frac{1}{2}\Gamma_a$, we obtain

$$-\frac{2\,\mathrm{Im}\,x_+}{2\pi} = \frac{1}{\hbar}|\langle c|D|g\rangle|^2 \frac{(q+\varepsilon)^2}{1+\varepsilon^2} \qquad (2.36)$$

which is the quantity $(1/\hbar)|\langle \Psi_{\bar{E}}|D|g\rangle|^2$ in terms of which autoionization is usually described, as discussed in Section II.A.

The other extreme case, strong field in the sense $|\alpha| \gg \Gamma_a, \gamma_g, \delta$, is also easily obtained. The square root can now be approximated by 2α and the roots by

$$x_+ = -\tfrac{1}{2}[\delta + (i/2)(\gamma_g + \Gamma_a)] \pm \alpha \qquad (2.37)$$

If, for the sake of illustration, we let $\gamma_g \to 0$ and $q \to \infty$, we obtain $x_\pm = -\frac{1}{2}[\delta + (i/2)\Gamma_a] \pm \tilde{\Omega}$ with the resulting expression

$$u_g(t) = \frac{1}{2\tilde{\Omega}}\left\{\left[\frac{1}{2}\left(\delta + \frac{i}{2}\Gamma_a\right) + \tilde{\Omega}\right]e^{-i(\tilde{\Omega} - \delta/2)t} - \left[\frac{1}{2}\left(\delta + \frac{i}{2}\Gamma_a\right) - \tilde{\Omega}\right]e^{i(\tilde{\Omega} + \delta/2)t}\right\}$$

$$(2.38)$$

which corresponds to the oscillatory behavior one expects from a level coupled strongly to another level. With the approximations $\gamma_g \to 0$ and $q \to \infty$ we made above, the equations have in fact been reduced to representing the coupling of $|g\rangle$ to another discreet state $|a\rangle$, which can decay (not back to $|g\rangle$) with a rate $\frac{1}{2}\Gamma_a$. Our approximations have all but eliminated the continuum which has only served in providing the decay Γ_a, but all interference has been taken out by setting $q \to \infty$. The frequency of oscillation is $\tilde{\Omega}$, (usually called Rabi frequency) and consistent with the approximations, $\tilde{\Omega}$ now is simply equal to D_{ag}, as it should if we only have two discreet levels coupled by a dipole transition.

Let us now consider the limit of strong field, but without the other approximations. Strictly speaking, strong field should be here defined by the inequality $|\tilde{\Omega}| \gg \Gamma_a$, which implies that the strength of the coupling between $|a\rangle$ and $|g\rangle$ becomes comparable to or larger than the coupling of $|a\rangle$ to the continuum. Indeed, Γ_a does not depend on the field while $\tilde{\Omega}$ is proportional to ε and it is through $\tilde{\Omega}$ that the field will affect the process of autoionization. But since γ_g is also field-dependent, it can not be arbitrarily ignored. Only if $q \to \infty$ can we ignore it, as we did above, because the interference is negligible in that case. For arbitrary q, but still strong field, we must keep γ_g and use Eq. (2.37) for x_+. The resulting expression is similar to Eq. (2.38), except that now instead of $\tilde{\Omega}$, we have $\alpha = \tilde{\Omega}(1 - i/q)$ in the exponentials. Formally, we still have the previous oscillatory behavior, but the presence of the quantity i/q indicates that the usual picture of a two-level system will appear altered because of the interference that q reflects. The features of this case are best seen in a numerical calculation of the line shape.

Before discussing the results of such calculations, let us recall that only in the weak-field case do we except the process to be describable by a transition probability per unit time. In most other cases, the underlying oscillatory behavior requires the calculation of the time-dependent quantity $P(T)$ of Eq. (2.23). It is possible that even for fields that are not weak, the oscillatory behavior is damped out for times larger than $1/\tilde{\Omega}$, in which case an effective transition probability per unit time is still valid. But still the line shape will be changed. More important, however, is the effect of the interaction time T on the line shape and this effect is important for strong field. Its physical origin is readily understood. If the field is sufficiently strong, the ionization can be 100% at the center of the resonance but also somewhat away from resonance. In fact, as the field becomes stronger, total ionization can occur at larger detunings. On the other hand, for a given field (even weak) ionization can be total if the time T is sufficiently large. This question is usually irrelevant in weak-field spectroscopy because the combination of intensities and times is such that justifies the simple transition probability interpretation completely. This ceases to be the case when lasers of powers such as 10^9 W cm^{-2} and of durations even as small as 10 ns are involved. It is not only in autoionization that this is important but also in any ionization or dissociation process with a relatively strong laser. Thus what time T one uses in the calculation or the experiment is an essential element in the final result.

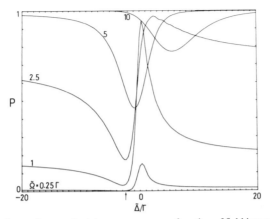

Fig. 1. Line shape of an autoionizing resonance as a function of field intensity. The dynamic detuning is represented by $\tilde{\Delta}$ and the position of the field-free minimum is indicated by the arrow. The interaction time is $T = 5\Gamma^{-1}$ and $q = 5$.

Let us consider now the effect of the field strength on the process of autoionization. The results of representative calculations are shown in Figs. 1 and 2. The time employed in the calculation of Fig. 1 is $T = 5\Gamma^{-1}$, which is sufficiently long (but not too long) for the typical profile to be obtained for weak intensity. The parameter q has been chosen equal to 5, which in weak-field autoionization gives rise to a significantly asymmetric profile, as is also evident in the bottom curve of Fig. 1. The figure contains five curves corresponding to various field intensities parameterized by $\tilde{\Omega}$ expressed in

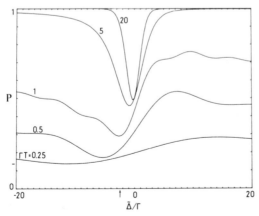

Fig. 2. Line shape of an autoionizing resonance as a function of interaction time for relatively strong field, i.e., $\tilde{\Omega} = 5\Gamma$. The arrow indicates the position of the field-free minimum and $q = 5$.

units of Γ, the autoionization width. The strongest field employed in the calculation corresponds to $\tilde{\Omega} = 10\Gamma$, which roughly speaking corresponds to a field-induced coupling between $|g\rangle$ and $|a\rangle$, about 10 times stronger than the coupling of $|a\rangle$ to the continuum. Clearly there is significant distortion of the field-free line shape. Recall that when two discreet states are coupled strongly by a field, there is broadening (the so-called power broadening), which however does not change the line qualitatively. Here, because of the presence of the continuum to which $|a\rangle$ is coupled by V and $|g\rangle$ by the field, we have an altogether different situation with qualitative changes of the line shape. The relative change of the line shape with intensity depends of course on q. In the limit $q \to \infty$, we recapture the case of simple power broadening corresponding to the strong coupling of two discrete states one of which decays via an interaction independent of the field.

Figure 2 represents the results of a calculation similar to that of Fig. 1, except that now the field intensity is kept fixed at the relatively strong value $\tilde{\Omega} = 5\Gamma$, while the line shape is calculated for different interaction times ranging from $T = 0.25\Gamma^{-1}$ (i.e., smaller than the autoionization lifetime) to $T = 20\Gamma^{-1}$ (much larger than the autoionization lifetime). The point to be made with this figure is that the interaction time is as important in determining the line shape as is the field intensity. In fact, the field intensity alone without the interaction time gives very little information about the possible change of the line shape. It should be noted here that the curve for the longest time ($\Gamma T = 20$) shows that ionization is 100% for detunings as large as 20 times the autoionization width, except around the minimum where ionization is much less efficient owing to the destructive interference between the channels $g \to a$, and $g \to c$; remember $q = 5$ in this figure as well. The same feature is also seen in Fig. 1. In both figures, the minimum is clearly shifting with changing intensity and/or time. Also to be noted in Fig. 2 are the oscillations apparent in two of the curves, but especially in the curve for $\Gamma T = 1$. They are the result of Rabi oscillations between g and a and are here manifested in the curve as a function of detuning as they would also be present in a curve of ionization as a function of time at fixed detuning. Evidently, it takes particular combinations of parameters for such oscillations to be discernible in the line shape.

It is now evident that neither the width of the line profile nor the position of its minimum are related simply to the atomic parameters q and Γ_a as they are under weak-field conditions. Two new quantities, field strength and time, play an equally determining role. Needless to say that if Γ_a and q are the desired quantities, the experiment must fulfill the weak-field conditions, which are easily achieved when a single-photon transition is possible. In a multiphoton transition, however, the field can not be weak and more often than not one will have to deal with the more complicated analysis.

III. AUTOIONIZING STATES IN MULTIPHOTON TRANSITIONS

A. Two-Photon Transition and Double Resonance

The two-photon transition is the simplest of all multiphoton transitions and we devote a separate subsection to its discussion in order to introduce certain generalizations which are easily extended to higher-order processes. To begin in fact with the simplest two-photon case, we consider first a resonant two-photon transition from an initial bound state $|g\rangle$ to an autoionizing state $|b\rangle$ via another (intermediate) state $|a\rangle$. We do, however, introduce a generalization by assuming $|a\rangle$ to also be an autoionizing state. Thus we have the option of studying the coupling of two autoionizing resonances by a laser field and at the same time reducing it by the appropriate choice of parameters, to the case of a two-photon transition from a bound to an autoionizing state via an intermediate bound state. Both are situations met in some of the recent experiments [2,3,7].

Since both $|a\rangle$ and $|b\rangle$ are autoionizing, they are coupled through V to continua, which will be denoted by $|c_1\rangle$ and $|c_2\rangle$, respectively. We assume two lasers (1) and (2) with photon occupation numbers and frequencies n_1, ω_1, n_2, ω_2. Laser (1) couples g with a and with c_1, while laser (2) couples a with b and with c_2. In addition, both lasers will, in general, couple a with c_2, b with c_1, as well as c_1 with c_2. States g and b are not coupled directly, usually for reasons of parity. For the same reason g is not coupled with c_2. Of course the choice of frequencies, namely $\hbar\omega_1 + \hbar\omega_2 \simeq E_b - E_g$, in the present context will not allow the coupling of g with b and c_2. The schematic representation of the various couplings defined above is shown in Fig. 3.

In the notation of Section II, the states of interest for the whole system now are:

$$|g'\rangle \equiv |g; n_1, n_2\rangle, \qquad |a'\rangle \equiv |a; n_1 - 1, n_2\rangle, \qquad |b'\rangle \equiv |b; n_1 - 1, n_2 - 1\rangle$$
$$|c_1'\rangle \equiv |c_1; n_1 - 1, n_2\rangle, \qquad |c_2'\rangle \equiv |c_2; n_1 - 1, n_2 - 1\rangle$$

Their energies can be taken as

$$E_g' = E_g + \hbar\omega_1 + \hbar\omega_2, \qquad E_a' = E_a + \hbar\omega_2,$$
$$E_b' = E_b, E_{c_1}' = E_{c_1} + \hbar\omega_2, \qquad E_{c_2}' = E_{c_2}$$

where we have again made use of the fact that only energy differences matter and have subtracted off the common quantity $(n_1 - 1)\hbar\omega_1 + (n_2 - 1)\hbar\omega_2$. On the basis of the couplings defined above and shown in Fig. 3, the following matrix elements are nonvanishing: $\langle g'|H^0|g'\rangle = E_g'$, $\langle a'|H^0|a'\rangle = E_a'$, $\langle b'|H^0|b'\rangle = E_b'$, V_{c_1a}, D_{ag}, V_{c_2b}, D_{ba}, D_{c_1g}, D_{c_2a}, D_{c_1b}, and $D_{c_1c_2}$. The

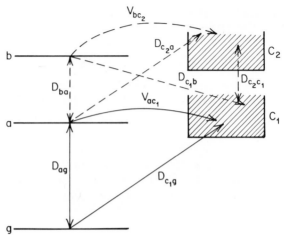

Fig. 3. Schematic representation of autoionizing states coupled to electromagnetic fields. D and V denote electromagnetic and configuration interactions, respectively. The shaded areas represent the continua to which the bound states a and b are coupled.

equations for the matrix elements of G that pertain to the states of interest can now be written straightforwardly using the procedure of Section II. They are

$$(z - E_g')G_g - D_{ga}G_a - \int dE_{c_1}' D_{gc_1}G_{c_1} = 1 \quad (3.1a)$$

$$-D_{ag}G_g + (z - E_a')G_a - \int dE_{c_1}' V_{ac_1}G_{c_1} - D_{ab}G_b - \int dE_{c_2}' D_{ac_2}G_{c_2} = 0 \quad (3.1b)$$

$$-D_{ba}G_a + (z - E_b')G_b - \int dE_{c_1}' D_{bc_1} - \int dE_{c_1}' V_{bc_2}G_{c_2} = 0 \quad (3.1c)$$

$$-D_{c_1g}G_g - V_{c_1a}G_a - D_{c_1b}G_b + (z - E_{c_1}')G_{c_1} - \int dE_{c_2}' D_{c_1c_2}G_{c_2} = 0 \quad (3.1d)$$

$$-D_{c_2a}G_a - V_{c_2b}G_b + (z - E_{c_2}')G_{c_2} - \int dE_{c_1}' D_{c_2c_1} = 0 \quad (3.1e)$$

These are the most general equations for this problem of double resonance. Because of the coupling $D_{c_2c_1}$, the two continua are coupled and can not be eliminated as in Section II. One way to proceed is to first diagonalize the continua through, for example, a K-matrix formalism as discussed most recently by Armstrong *et al.* [16] in a context related to ours. The resulting equations do give some general formal idea of the interplay between the various couplings. But they are rather cumbersome. We neglect this term here confining ourselves to situations in which its contribution is small, as is expected to be the case when a substantial portion of the transition strength is in the autoionizing resonance. This will by no means be generally true, but it is too early in the state of the field to tell exactly how and when it will be significant. The extreme opposite case occurs in the above-threshold absorptions in rare gases when $D_{c_2c_1}$ would be the only contributing term. But we know of no contributing autoionizing states there. As we have shown in detail elsewhere [15], G_{c_1} and G_{c_2} can be decoupled without necessarily setting $D_{c_2c_1} = 0$ at this stage. It can be done later in the calculation in a way that retains part of the contribution of $D_{c_2c_1}$.

Proceeding from here on with the approximation $D_{c_2c_1} = 0$ and after some algebraic manipulations analogous to those of Section II, we arrive at the following set of equations:

$$[x + (i/2)\gamma_g]G_g - \bar{D}_{ga}G_a - \overline{D_{gb}^{(2)}}G_b = 1 \quad (3.2a)$$

$$-\bar{D}_{ag}G_g + [x + \delta_1 + (i/2)\gamma_a + (i/2)\Gamma_a]G_a - \bar{D}_{ab}G_b = 0 \quad (3.2b)$$

$$-\overline{D_{bg}^{(2)}}G_g - \bar{D}_{ba}G_a + [x + \delta_1 + \delta_2 + (i/2)\gamma_b + (i/2)\Gamma_b]G_b = 0 \quad (3.2c)$$

with the definitions:

$$\bar{D}_{ga} = D_{ga} + \mathbb{P}\int dE'_{c_1} \frac{D_{gc_1}V_{c_1a}}{E'_g - E'_{c_1}} - i\pi(D_{gc_1}V_{c_1a})_{E'_g} \quad (3.3a)$$

$$\overline{D_{gb}^{(2)}} = \mathbb{P}\int dE'_{c_1} \frac{D_{gc_1}D_{c_1b}}{E'_g - E'_{c_1}} - i\pi(D_{gc_1}D_{c_1b})_{E'_g} \quad (3.3b)$$

$$\bar{D}_{ab} = D_{ab} + \mathbb{P}\int dE'_{c_1} \frac{V_{ac_1}D_{c_1b}}{E'_g - E'_{c_1}} \, i\pi(V_{ac_1}D_{c_1b})_{E'_g}$$
$$+ \mathbb{P}\int dE'_{c_2} \frac{D_{ac_2}V_{c_2b}}{E'_g - E'_{c_2}} - i\pi(D_{ac_2}V_{c_2b})_{E'_g} \quad (3.3c)$$

$$S_g + \frac{i}{2}\gamma_g \equiv \mathbb{P}\int dE'_{c_1} \frac{|D_{c_1g}|^2}{E'_g - E'_{c_1}} + i\pi|D_{c_1g}|^2_{E'_g} \quad (3.4a)$$

$$S_a + \frac{i}{2}\gamma_a \equiv \mathbb{P}\int dE'_{c_2} \frac{|D_{c_1g}|^2}{E'_g - E'_{c_2}} + i\pi|D_{c_2a}|^2_{E'_g} \quad (3.4b)$$

$$S_b + \frac{i}{2}\gamma_b \equiv \mathbb{P}\int dE'_{c_1} \frac{|D_{c_1b}|^2}{E'_g - E'_{c_1}} + i\pi|D_{c_1b}|^2_{E'_g} \quad (3.4c)$$

$$F_a + \frac{i}{2}\Gamma_a \equiv \mathbb{P}\int dE'_{c_1} \frac{|V_{c_1a}|^2}{E'_g - E'_{c_1}} + \pi|V_{c_1a}|^2_{E'_g} \quad (3.4d)$$

$$F_b + \frac{i}{2}\Gamma_b \equiv \mathbb{P}\int dE'_{c_2} \frac{|V_{c_2b}|^2}{E'_g - E'_{c_2}} + \pi|V_{c_2b}|^2_{E'_g} \quad (3.4e)$$

The interpretation of these shifts and widths are analogous to those of Section II. To obtain $\bar{D}_{ag}$ from $\bar{D}_{ga}$, we simply transpose the indices in all matrix elements but do *not* take the complex conjugate of the whole expression. The same holds true for $\bar{D}_{ba}$ and $D_{bg}^{(2)}$. The detunings

$$\delta_1 \equiv \hbar\omega_1 - (E_a + F_a + S_a) + (E_g + S_g) \tag{3.5a}$$

$$\delta_2 \equiv \hbar\omega_2 - (E_b + F_b + S_b) + (E_a + F_a + S_a) \tag{3.5b}$$

are as before dynamic detunings which incorporate the ac-Stark shifts.

The total ionization is given by

$$P(t) = 1 - |U_{gg}(T)|^2 - |U_{ag}(T)|^2 e^{-\Gamma_a(t-T)} - |U_{bg}(T)|^2 e^{-\Gamma_b(t-T)} \tag{3.6a}$$

when $t \geq T$ (see relevant remarks in Section II). In most realistic situations $\Gamma(t - T) \gg 1$, in which case

$$P = 1 - |U_{gg}(T)|^2 \tag{3.6b}$$

which is equal to

$$\sum_{j=1}^{2} \int dE_{c_j} |U_{c_jg}(T)|^2$$

A more familiar case is now easily obtained as a special case of Eqs. (3.2). Assume that $|a\rangle$ is a bound state below the continuum threshold (or even a narrow, very long lived state compared to autoionizing state $|b\rangle$). Then we can neglect the continuum $|c_1\rangle$ and as a consequence $D_{gb}^{(2)}$, γ_g, S_g, F_a, Γ_a, and γ can all be taken equal to zero. The resulting equations then are

$$xG_g - \Omega_1 G_g = 1 \tag{3.7a}$$

$$-\Omega_1 G_g + [x + \delta_1 + (i/2)\gamma_a]G_a - \tilde{\Omega}_2[1 - (i/q_b)]G_b = 0 \tag{3.7b}$$

$$-\tilde{\Omega}_2[1 - (i/q_b)]G_a + [z + \delta_1 + \delta_2 + (i/2)\Gamma_b]G_b = 0 \tag{3.7c}$$

where in accordance with previous notation

$$\Omega_1 \equiv D_{ag} \tag{3.8a}$$

and

$$\bar{D}_{ba} \equiv D_{ba} + \mathbb{P}\int dE'_{c_2}\frac{V_{bc_c}D_{c_2a}}{E'_g - E'_{c_2}} - i\pi(V_{bc_2}D_{c_2a})_{E'_g} \equiv \tilde{\Omega}_2[1 - (i/q_b)] \tag{3.8b}$$

Now Ω_1 is a conventional Rabi frequency, while $\tilde{\Omega}_2$ is an effective Rabi frequency of the sort we encountered in Section II. We have only one q parameter referring to the single autoionizing resonance $|b\rangle$ and its continuum $|c_2\rangle$ that have remained in the problem. Note that we could have kept Γ_a in the problem, without contradiction, if we interpreted it as the decay width of state $|a\rangle$.

The physical situation described by Eqs. (3.7) is a somewhat generalized double resonance. It is more general in that it involves an autoionizing state and not three bound states as is the traditional double optical resonance (DOR) situation. It also has something in common with another version of DOR, namely DOR with ionization of the upper state, as the observed quantity. In fact it is equivalent to that case, except that here the ionization of $|b\rangle$ (upper state) does not require a field but is caused by intra-atomic interactions. As an additional consequence of that interaction, we have the interference represented by q_b, which is absent from both of the above comparable situations. Letting $q_b \to \infty$ and reinterpreting Γ_b appropriately makes the correspondence exact.

The general solutions of Eqs. (3.2) and (3.7) are expressed as linear combinations of oscillating functions involving the roots of a cubic equation resulting from the determinant of the coefficients of the G. We do not explore here the general solutions of Eqs. (3.2) because it is outside the scope of this article. Also, not much is known yet experimentally about this general physical situation depicted in Fig. 3. It just is too early in the development of the field. It is, however, worth devoting some space to Eqs. (3.7) and their implications since there already exists some literature [15–19] on the subject, albeit theoretical.

We explore now a special case of Eqs. (3.7), namely the so-called weak-probe case. We assume that the field connecting $|g\rangle$ and $|a\rangle$ is weak, in the sense that $\tilde{\Omega}_1 \ll \gamma_a$, Γ_b, $\tilde{\Omega}_2$, and we explore the profile of total ionization as a function of ω_1 (or equivalently δ_1) for various values of the second field and hence of $\tilde{\Omega}_2$. Let us first note that, under this weak-probe condition, a transition probability for total ionization can be derived analytically. It is given by

$$\frac{dP}{dt} = \Omega_1^2 \gamma_a \frac{\delta_1 + \delta_2 + \frac{1}{2}\Gamma_b q_b}{|f(\delta_1)|^2} \quad (3.9a)$$

where

$$f(\delta_1) = [\delta_1 + (i/2)\gamma_a][\delta_1 + \delta_2 + (i/2)\Gamma_b] - \tilde{\Omega}_2^2[1 - (i/q_b)]^2. \quad (3.9b)$$

Obviously, the line profile as a function of δ_1 will, in general, have two peaks determined by the roots of $f(\delta_1)$. If $\tilde{\Omega}_2$ is also small, we will simply obtain a single peak positioned so that it conserves energy between initial and final state of the two-photon transition. Thus the position of this single peak will depend on δ_2. To fix matters for this discussion, let us take $\delta_2 = 0$ and explore what happens as $\tilde{\Omega}_2$ varies from weak to strong; strong here meaning $\tilde{\Omega}_2 > \Gamma_b$. If $|b\rangle$ were not an autoionizing state but simply a bound state, we would have the typical DOR situation, which would exhibit the well-known ac-Stark splitting; namely two peaks separated by $\sim 2\tilde{\Omega}_2$ when

$\tilde{\Omega}_2 > \Gamma_b$. The peaks would have equal heights when $\delta_2 = 0$, that is when the strong field is exactly on resonance with the unperturbed (zero-field) states $|a\rangle$ and $|b\rangle$. If $\delta_2 \neq 0$ the peaks have unequal heights.

One expects something similar to happen in the present case as well. But also some new features should appear owing to the presence of interference with the continuum. Thus whatever differences from the usual DOR exist, they should appear for q finite, while the profile should revert to the typical DOR for $q \to \infty$. One could write analytical expressions for the profile based on Eqs. (3.9). Keeping in mind, however, that even when a rate is applicable, the actual profile may depend on the interaction time; we give preference here to a graphical illustration shown in Fig. 4. The six different frames correspond to different values of $\tilde{\Omega}_2$ denoted simply by Ω in the figure. For $\Omega < \Gamma$ we have the expected single peak, while for $\Omega = \Gamma$ the two peaks of ac-Stark splitting begin to appear with the splitting increasing as Ω increases. Already we have a difference from usual DOR. The peaks have unequal heights even though the strong field is exactly resonant ($\delta_2 = 0$). It is the first manifestation of the interference, or from another viewpoint, the manifestation of the asymmetry in the weak-field autoionization profile of $|b\rangle$. A more remarkable feature appears as the field is increased ever further to about $\Omega \cong 5\Gamma$. One of the peaks narrows to a theoretically zero width, while the other broadens. As the intensity is increased even further both peaks reappear. Depending on one's favored way of interpreting such effects, there are various pictures one can construct. For example, we may note that there are two paths via which $|a\rangle$ can make

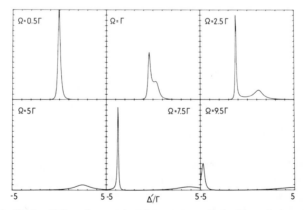

Fig. 4. Ac-Stark splitting of an autoionizing resonance in double optical resonance ($\Delta = 0$) and its strength is varied from $\Omega = 0.5\Gamma$ to 9.5Γ. The detuning Δ' of the weak (probe) field is measured in units of autoionization width Γ. The calculation has been performed in the limit of a steady state with $q = 5$. Stark shifts have been neglected.

a transition to the continuum: directly or via $|b\rangle$. In both cases a photon is absorbed, but in the path through $|b\rangle$ the interaction V is also involved. The strength of the field affects the relative magnitude of the transition amplitudes through these two paths which interfere as is obvious by simple inspection of Fig. 3. At some critical intensity, this interference is just right for the transition via one path to undo, so to speak, the one through the other path. In a sense, this means that a component of state $|b\rangle$ does not autoionize because the interfering path takes it from the continuum back to the discreet states. Another way of picturing it is obtained by noting that from the coupling of $|a\rangle$ and $|b\rangle$ two new states, linear superpositions of both, are obtained. The continuum is mixed with both and leads to decay, except at a particular intensity for which the coherent superposition of the continuum with the mixing of $|a\rangle$ and $|b\rangle$ causes one of the two mixed states to acquire zero width, thereby becoming stable against ionization. In yet another way of expressing the same effect, one could say that the Rabi oscillation between $|a\rangle$ and $|b\rangle$ "keeps" the system from leaking into the continuum.

The above effect of line narrowing has been rediscovered in a slight variation by Rzazewski and Eberly [18], who seem to not have realized that the effect they called "confluence of coherences" was mathematically and in physical content identical to the effect in one of the references [17] they site. When two discreet states $|g\rangle$ and $|a\rangle$ are coupled strongly by a near-resonant field, there will be ac-Stark splitting. If total ionization from $|a\rangle$ is observed, it is obvious that the splitting is not observable since the measured quantity is the total transition into a continuous spectrum. In other words, there is no probe that detects the splitting. It is equally obvious that if the kinetic energy spectrum of the photoelectrons is analyzed, it will exhibit the two peaks of the splitting because the energy analysis serves as probe. Thus if $|g\rangle$ is a ground state and $|a\rangle$ is an excited state, and a single laser is employed to couple $|g\rangle$ to $|a\rangle$ and also ionize $|a\rangle$, (if permitted energetically) only photoelectron energy analysis will reveal the splitting. Assume now that $|a\rangle$ is embedded in a continuum into which it autoionizes. No laser is necessary for the ionization. But the radiation coupling between $|g\rangle$ and $|a\rangle$ will cause splitting which again will be observed only if the kinetic energy of the ejected electrons is analyzed and not in the total ionization signal. This is the content of the Rzazewski and Eberly paper in which the width of one of the Stark peaks is found to vanish at a particular critical intensity, for the reasons discussed earlier in connection with the double resonance arrangement. The phenomenon is the same: the ac-Stark splitting of a transition coupling an autoionizing resonance with a bound state. Whether it is observed in double resonance or through the kinetic energy of the emitted electrons is simply a matter of how the splitting is probed.

Needless to add that even in the double resonance arrangement, the splitting could also be observed by leaving the frequency of the weak (probe) laser fixed and observing instead the kinetic energy of the emitted electrons. Not that this scheme is more desirable, but we point it out to underscore the total equivalence between the two schemes. It is of course because of this equivalence that the vanishing of the width occurs at the same critical intensity in both cases. Precise predictions for this critical intensity, however, ought to be taken with caution because they depend somewhat on the model and the approximations made in the calculation.

Assuming that one decides to attempt the observation of the ac-Stark splitting in this context, a simple numerical example is useful in revealing an important experimental difference between the above two schemes. Let us take an autoionization width of about 20 cm^{-1}, which is typical of the autoionizing states observed in multiphoton spectroscopy. For the splitting to be observable, it has to be at least three to five times larger than the width, which means about 100 cm^{-1}. The resolution necessary for the scanning of this splitting would then be $\sim 10\ cm^{-1}$, which corresponds to a resolution of about 1 meV in photoelectron kinetic energy. This is a resolution that requires rather complicated instrumentation. It is a difficult experiment. In DOR, on the other hand, a resolution of 10 cm^{-1} or two orders of magnitude less poses no problem because it is a resolution referring to an optical or uv photon frequency. This makes it an extremely easy experiment from that point of view. It could be argued of course that the problem of resolution in observing electron kinetic energy can be alleviated by choosing a broad resonance. This, however, necessitates larger splittings and consequently larger intensities, which in turn introduces higher-order transitions upward (ionization), thus seriously limiting, if not invalidating, the two- or three-level systems on which the model is based. It is worth recalling that the splittings so far observed in bound–bound transitions have been of the order of a few GHz. From the experimental point of view, therefore, there is an enormous difference between the above two ways of detecting ac-Stark splitting in autoionization.

Irrespective of the relative merits of these two ways of observation, we should not neglect posing the question: Why should one attempt the observation at all and what would one learn? The observation of the ac-Stark splitting itself is probably uninteresting as it has little to add to the numerous existing studies [20] of the same effect between bound states; a topic that has received much attention in connection with resonance fluorescence and double optical resonance under strong (saturating) fields. The interaction with the continuum does introduce a new twist, but it is something that is readily understood theoretically. An experimental demonstration would be little more than the implementation of a cute effect.

Moreover, if it is attempted, it will probably have to cope with possible saturation owing to the interaction time discussed earlier; not a prohibitive but nevertheless an annoying complication.

Although we would not urge experimentalists to set out searching for the effect at all costs, it will eventually prove relevant to the study of auto-ionizing states under strong fields. It certainly represents an unusual line-distortion effect and will have to be accounted for in the extraction of atomic parameters from such experiments. In addition, the narrowing of one of the peaks does correspond to the increase of the lifetime of one of the dressed states and may eventually prove useful in processes having to do with the generation of radiation through the involvement of autoionizing states. For the moment, it is an interference effect that one must keep in mind.

There are several ways in which this interference can be manifested in more complicated situations. For example, returning to Fig. 3, one may ponder whether the narrowing would occur if both $|a\rangle$ and $|b\rangle$ autoionize. The question is intriguing because both states are then unstable due to their coupling to the continua, and the vanishing of the width would imply the appearance of a stable component (state) out of the unstable structure spanned by $|a\rangle$, $|b\rangle$, and the continua. It can be shown that this is indeed the case even under the most general conditions in which all couplings depicted in Fig. 3 are nonvanishing. Another related aspect that emerges out of this analysis is that two autoionizing states coupled to the same con-tinuum and spaced by less than the sum of their widths (overlapping) can separate at a critical intensity that couples them to a bound state. The resulting profile will exhibit two peaks, thus revealing the structure under-lying the broad profile of the two overlapping resonances. We do not elabo-rate here on this and other similar effects. Undoubtedly they will eventually be found relevant to the multiphoton laser spectroscopy of autoionizing resonances, but it is somewhat early to predict the precise context.

B. Higher-Order Transitions

Having sampled the types of new effects that appear in single- and two-photon excitation of an autoionizing resonance by a strong field, we turn now to a more common type of multiphoton excitation: The nonresonant multiphoton excitation. Since we always have in mind an autoionizing state and wish to explore, among other features, its line shape under multiphoton excitation, by nonresonant we mean absence of resonances with intermediate states. We will always assume that we are tuning around the autoionizing resonance and are therefore around resonance with the autoionizing state itself.

Beginning with the two-photon case, we can easily obtain its mathematical description by generalizing the equations of Section III.A. Consider Eqs. (3.1) and assume an infinite set of intermediate states $|a\rangle$, and one autoionizing state $|b\rangle$ coupled to a single continuum $|c\rangle$ through the matrix element V_{cb}. The evidently resulting set of equations is

$$(z - E'_g)G_g - \sum_a D_{ga}G_a = 1 \qquad (3.10a)$$

$$-D_{ag}G_g + (z - E'_a)G_a - D_{ab}G_b - \int dE'_c D_{ac}G_c = 0 \qquad (3.10b)$$

$$-\sum_a D_{ba}G_a + (z - E'_b)G_b - \int dE'_c V_{bc}G_c = 0 \qquad (3.10c)$$

$$-\sum_a D_{ca}G_a - V_{cb}G_b + (z - E'_c)G_c = 0 \qquad (3.10d)$$

None of the intermediate states $|a\rangle$ are near resonant with the energy $E_g + \hbar\omega$. On the other hand, $E_g + 2\hbar\omega$ is assumed to vary around E_b. The reader familiar with two-photon ionization will recognize that G_a and G_c must be eliminated from these equations. Usually the intermediate states are eliminated first leading to an effective two-photon matrix element coupling $|g\rangle$ with the continuum, which is in turn eliminated, its effect being expressed by the ionization widths and shifts. Here we have an additional interaction, namely V_{bc}, to cope with. Some care is necessary in performing such eliminations in order to avoid missing the resulting interference terms. It is necessary to ensure that the final result is independent of the order of elimination of the intermediate states and of the continuum, within the same approximations, which are the pole approximation in eliminating the continuum, and the nonresonance approximation, i.e., $E_a - E_g \neq \hbar\omega$ for all E_a. The proof of this statement is not trivial although its physical content is obvious. We do not give the proof here because it can be found elsewhere [21].

After the above eliminations, we expect to be left with two equations governing the evolution of $|g\rangle$ and $|a\rangle$. They are

$$\left(z - E_g - 2\hbar\omega - \int dE'_c \frac{|D^{(2)}_{cg}|^2}{z - E'_c}\right)G_g - \left(D^{(2)}_{gb} + \int dE'_c \frac{D^{(2)}_{gc}\mathscr{U}_{cb}}{z - E'_c}\right)G_b = 1 \quad (3.11a)$$

$$-\left(D^{(2)}_{bg} + \int dE'_c \frac{\mathscr{U}_{bc}D^{(2)}_{cg}}{z - E'_c}\right)G_g + \left(z - E_b - \int dE'_c \frac{|\mathscr{U}_{cb}|^2}{z - E'_c}\right)G_b = 0 \quad (3.11b)$$

which formally are identical to Eqs. (2.15) describing the coupling of an autoionizing state to an initial state by a single-photon transition. The difference between Eqs. (2.15) and (3.11) lies in the different meaning of the matrix elements. Now we have second-order (two-photon) effective matrix elements coupling $|g\rangle$ to $|b\rangle$ and the continuum. Thus

$$D_{cg}^{(2)} \equiv \sum_{a'} \frac{D_{c'a'} D_{a'g'}}{E_{g'} - E_{a'}} = \mathscr{E}^2 \sum_a \frac{\mu_{ca}\mu_{ag}}{E_g - E_a + \hbar\omega} \tag{3.12a}$$

is an effective two-photon matrix element coupling the initial state to the continuum. Similarly,

$$D_{bg}^{(2)} \equiv \sum_{a'} \frac{D_{b'a'} D_{a'g'}}{E_{g'} - E_{a'}} = \mathscr{E}^2 \sum_a \frac{\mu_{ba}\mu_{ag}}{E_g - E_a + \hbar\omega} \tag{3.12b}$$

is an effective 2-photon matrix element coupling $|g\rangle$ with the autoionizing state $|b\rangle$. A more profound difference is reflected in $\mathscr{U}_{cb}$, which is defined by

$$\mathscr{U}_{cb} \equiv V_{cb} + D_{c'b'}^{(2)} = V_{cb} + \mathscr{E}^2 \sum_a \frac{\mu_{ca}\mu_{ab}}{E_g - E_a + \hbar\omega} \tag{3.13}$$

where clearly $D_{cb}^{(2)}$ is an effective two-photon matrix element coupling the continuum with $|b\rangle$, which is also coupled to the continuum by V_{cb}. The two couplings interfere, the net result of this interference being intensity-dependent since V_{cb} is independent of the intensity, while $D_{c'b'}^{(2)}$ is proportional to the square of the strength of the electric field. The quantity $\mathscr{U}_{bc}$ appearing in Eqs. (3.11) emerges out of the systematic elimination of the intermediate states and of the continuum, and it enters Eqs. (3.11) in exactly the same way that the configuration interaction V_{bc} enters Eqs. (2.15). With the aid of the schematic representation shown in Fig. 5 we can now construct a picture of two-photon (nonresonant) autoionization as it emerges from Eqs. (3.11): the initial state $|g\rangle$ is coupled directly to the continuum by a two-photon transition, it is coupled to the autoionizing state $|b\rangle$ by a two-photon transition, while $|b\rangle$ is coupled to the continuum by the configuration interaction V_{cb} modified by an interfering two-photon transition between $|b\rangle$ and $|c\rangle$. The latter is represented by the dashed line of Fig. 5 and constitutes a drastically new feature of two-photon autoionization as compared to its

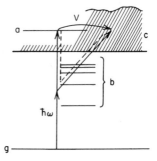

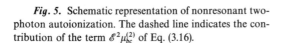

Fig. 5. Schematic representation of nonresonant two-photon autoionization. The dashed line indicates the contribution of the term $\mathscr{E}^2\mu_{bc}^{(2)}$ of Eq. (3.16).

single-photon counterpart. Before discussing the physical implications of the above picture further, it will prove useful to derive a few more formal results.

In view of the formal similarity between Eqs. (2.15) and Eqs. (3.11), the same steps that lead to the weak-field limit of Eqs. (2.15) can be employed for the same limit of Eqs. (3.11). Weak-field limit now corresponds to a different range of absolute intensities. It has to be sufficiently strong for the two-photon transitions to be significant, but not so strong as to saturate the transition $|g\rangle \to |b\rangle$. The resulting expressions for the transition probability per unit time is

$$W_2 = 2\pi I^2 |\mu_{cg}^{(2)}|^2 \frac{(\tilde{q}_2 + \varepsilon)^2}{1 + \varepsilon^2} \tag{3.14}$$

where I is the light intensity, which is proportional to $|\mathscr{E}|^2$,

$$\mu_{cg}^{(2)} \equiv \sum \frac{\mu_{ca}\mu_{ag}}{E_g - E_a + \hbar\omega}$$

and $\tilde{q}_2$ is defined by

$$\tilde{q} \equiv \frac{\mu_{bg}^{(2)} + \mathbb{P} \int dE_c [\mathscr{U}_{bc}\mu_{cg}^{(2)}/(\tilde{E} - E_c)]}{\pi[(V_{bc} + \mathscr{E}^2 \mu_{bc}^{(2)})\mu_{cg}^{(2)}]_{(E_c = \tilde{E})}} \tag{3.15}$$

We also introduce the quantity

$$\tfrac{1}{2}\tilde{\Gamma}_b \equiv \pi|V_{bc} + \mathscr{E}^2 \mu_{bc}^{(2)}|^2 = \pi|\mathscr{U}_{bc}|^2 \tag{3.16}$$

in terms of which the detuning ε is defined by

$$\varepsilon \equiv [2\hbar\omega - (\tilde{E}_b - E_g)]\tfrac{1}{2}\tilde{\Gamma}_b \tag{3.17}$$

where $\bar{E}_b \equiv E_b + F_b$ is the energy of $|b\rangle$ modified by the usual shift F_b due to the configuration interaction. The parameters $\tilde{q}_2$ and $\tilde{\Gamma}_b$ and the counterparts of q and Γ_a defined in Section II. Note, however, a very important difference: whereas q and Γ_a are atomic parameters independent of the radiation, $\tilde{q}$ and $\tilde{\Gamma}_b$ do depend on the intensity of the radiation.

This dependence on the intensity has significant consequences on the process. First, $\tilde{\Gamma}_b$ is no longer the autoionization width but is still the apparent width of the observed resonance, as is obvious from the above equations. It follows then that the width of the resonance will change with intensity in a way that depends on the relative sign of V_{bc} and $\mu_{bc}^{(2)}$. As long as the intensity is sufficiently low for V_{bc} to dominate, $\tilde{\Gamma}_b$ will be the width of the bare resonance determined by configuration interaction. With increasing intensity, and as $\mathscr{E}^2 \mu_{bc}^{(2)}$ becomes comparable to V_{bc}, the width $\tilde{\Gamma}_b$ will decrease or increase depending on whether the two terms have opposite or the same signs, respectively. With further increase in the intensity, of course $\tilde{\Gamma}_b$ will eventually increase becoming dominated by the term $\mathscr{E}^2 \mu_{bc}^{(2)}$. The dependence of $\tilde{q}_2$ on intensity is less straightforward. It still comes from the intensity-dependent term in $\mathscr{U}_{bc}$ and reverts to the intensity-independent parameters q for low intensity. Whether it increases or decreases with intensity will also depend on the other quantities appearing in Eq. (3.15). We have thus found that the width and shape of the resonance will change with intensity in a rather complicated manner including the possibility of narrowing at some particular range of intensities. An illustration of this effect will be discussed later. For now, let us note that this narrowing appears in a context different from the previously discussed narrowing in double resonance.

The generalization of the previous results to three-photon ionization is somewhat lengthier but is based on the same ideas. We have two infinite sets of intermediate states to eliminate in addition to the continuum. The resulting three-photon effective matrix elements contain in this case double summations over intermediate states, which are of course assumed to be nonresonant with the photon frequency. The complete three-photon transition to the autoionizing state is understood here to involve a single laser. This laser is assumed to be sufficiently strong for the three-photon process to be observable, but not so strong as to saturate the three-photon transition to the autoionizing resonance. On the basis of existing experimental evidence [3] and theoretical estimates [22], these intensities lie in the region of 10^8–10^{11} W cm^{-2} for near infrared and optical frequencies. Under such conditions, a transition probability per unit time is valid. In its derivation care must be taken to account for the interference with the process of autoionization (configuration interaction). Without going into the algebraic details of the derivation, we quote here the end result whose plausibility should be easy to accept after the previous derivation for the two-photon case.

The transition probability per unit time (rate) is again expressed as

$$W_3 = 2\pi I^3 |\mu_{cg}^{(3)}|^2 (\tilde{q}_3 + \varepsilon)^2/(1 + \varepsilon^2) \tag{3.18}$$

where $\mu_{cg}^{(3)}$ is a three-photon effective matrix element between the continuum $|c\rangle$ and the initial state $|g\rangle$. The parameter $\tilde{q}_3$ is given by

$$\tilde{q}_3 = \frac{\mu_{bg}^{(3)} + \mathbb{P} \int dE_c \, \mathscr{U}_{bc} \mu_{cg}^{(3)}/(\tilde{E} - E_c)}{\pi[(V_{bc} + \mathscr{E}^2 \mu_{bc}^{(2)})\mu_{cg}^{(3)}]_{(E_c = \tilde{E})}} \tag{3.19}$$

where, as before $\mathscr{U}_{bc} = V_{bc} + \mathscr{E}^2 \mu_{bc}^{(2)}$, and the resonance width is given by

$$\tfrac{1}{2}\tilde{\Gamma}_b \equiv \pi|V_{bc} + \mathscr{E}^2 \mu_{bc}^{(2)}|^2 \tag{3.20}$$

The detuning is now expressed as

$$\varepsilon \equiv \frac{3\hbar\omega - (\bar{E}_b - E_g)}{\tfrac{1}{2}\tilde{\Gamma}_b} \tag{3.21}$$

The excitation energy is $\tilde{E} = E_g + 3\hbar\omega$. The modification of the width and of $\tilde{q}_3$ by the intensity occurs in exactly the same manner as in the two-photon case. This is owing to our having included in this calculation only the lowest-order (nonvanishing) modification of V_{bc}. In fact, this modification, represented by $\mathscr{E}^2\mu_{bc}^{(2)}$, will always be the lowest-order modification of V_{bc} for multiphoton autoionization of any order. This does not imply that it will always be the dominant term, although it will be so for low intensity. There are several interesting aspects of this question which need not be elaborated upon here since we do not go beyond three-photon ionization in this chapter.

The discussion that followed the equations for two-photon autoionization applies equally well to the three-photon process. We expect the same type of intensity effects on the width and shape of the resonance. In fact we may now view single-photon and multiphoton autoionization from a unified point of view, in the regime of "weak" intensities, where a transition probability per unit time is applicable. In general, we have a transition probability and hence a line profile expressed in the form $(\tilde{q} + \varepsilon)^2/(1 + \varepsilon^2)$ multiplied by a matrix element connecting $|g\rangle$ with the continuum. For the single-photon case, this is a usual dipole matrix element, while for the multiphoton case, it is an effective matrix element of the appropriate order. The parameter $\tilde{q}$ and the width $\tilde{\Gamma}$ will in general be intensity-dependent quantities except for single-photon autoionization where they are atomic parameters independent of the radiation as long as the field is weak. The expressions for $\tilde{q}$ and $\tilde{\Gamma}$ are formally the same in all cases as long as the matrix elements of μ connecting $|g\rangle$ and $|b\rangle$ and the continuum are interpreted as effective matrix elements of the appropriate order and V_{bc} is replaced by $\mathscr{U}_{bc} = V_{bc} + \mathscr{E}^2\mu_{bc}^{(2)}$, which will always contain the two-photon matrix element connecting $|b\rangle$ to $|c\rangle$ via the last set of intermediate states. Deviations from the low-intensity profile appear when $\mathscr{E}^2\mu_{bc}^{(2)}$ becomes comparable to V_{bc}. Because of such deviations, we do not expect the profiles of multiphoton autoionization to always be the typical profiles nor do we expect to be able to extract correct atomic parameters from such profiles unless the possible intensity effects are taken into account correctly. The width of the resonance will eventually increase with radiation intensity but it may first undergo narrowing to (theoretically) zero width.

There is an important effect that has been left out of this derivation: the ac-Stark shift of the intermediate states relative to $|g\rangle$ and $|b\rangle$. We emphasize that we are referring to the Stark shift (not splitting) of nonresonant states which varies linearly with light intensity. Its immediate effect will be to bring closer to resonance some intermediate state, thereby enhancing $\mathscr{E}^2\mu_{bc}^{(2)}$, thus making $\tilde{\Gamma}_b$ more sensitive to intensity. Such shifts are automatically included in the basic formalism of this chapter and arise naturally in the process of elimination of virtual states, as we have seen explicitly in earlier parts of our discussion. To avoid mathematical complexity in our discussion we tacitly omitted their inclusion in the nonresonant processes. It is well-known and experimentally documented that ac-Stark shifts have profound effects on multiphoton ionization as is amply shown in other chapters of this volume.

IV. CONCLUDING REMARKS

In this chapter we have shown how to formulate autoionization in strong fields and have explored certain consequences of this formalism by explicitly deriving specific effects to be expected under various experimental contexts. The experimental situation is not fully formed yet and as of the writing of this chapter we must be content with some general remarks in relation to experiment.

There are no observations of the ac-Stark splitting of a transition to an autoionizing resonance. We are aware of attempts [23] at such observations which have only resulted in a broad profile without any detectable splitting. The conditions of these attempts (combination of intensity and pulse duration) seem to be compatible with the interpretation of "time broadening," which would mask the splitting if it were present. Given the lack of information on oscillator strengths of the autoionizing transitions involved in these experiments, one can not be sure the splitting was present, i.e., it exceeded the autoionization width. So far no one seems to have taken the trouble to design an experiment appropriate for the observation.

On the other hand, there are experimental results of multiphoton ionization of alkaline earths as we noted in the beginning. The most detailed set of data is contained in the paper by Feldmann and Welge, which is summarized in Chapter 9. A few aspects of their data have been compared [21] to calculations and there appears to exist some agreement even though the calculations were not sufficiently detailed to warrant more than qualitative agreement. Thus two of their angular distributions seem to behave as predicted by these calculations. Also calculations [21] of absolute cross sections for three-photon ionization of Sr and five-photon ionization of Sr^+ by the second harmonic of the Nd-YAG laser seem to be qualitatively compatible with the data. More than qualitative comparison is not feasible in any case since there are no direct measurements of absolute cross section. A more detailed calculation intended to test the intensity effects on the line profile has been published very recently [22]. It has to do with the autoionizing resonance of Sr which is identified as (5p6s) $^1P_1$ and which can be excited by a three-photon transition. Again, direct comparison with the

experiment is not feasible at this point since the experimental data appear
to involve the simultaneous excitation of at least one more resonance at
about the same wavelength. But the calculation can at least be viewed as
an illustration of the effect in a real atom. The results do indeed show a
narrowing of the resonance at an intensity 5.3×10^{11} W cm^{-2}. Since the
ac-Shark shifts were not included in that calculation, their order of magnitude suggests that the effect may actually occur at an intensity one order of
magnitude lower, if all other quantities remain the same. It is a small and
isolated piece of evidence, but it does seem to suggest that these effects may
have been involved in the unusual profiles observed in the experiments on
Sr.

   This is about all one can say for the moment with respect to comparisons
of theory with experiment. Most probably, the situation will change drastically before this chapter appears in print. Considerable theoretical work is
in progress and refinements of the experiments are either planned or under
way. It is through such refinements that we will hopefully learn, or at least
obtain clues, about the possibility of successive transitions between autoionizing states. It is through angular distributions that much can be learned
about the particular autoionizing states that have been observed in three-
photon transitions of Sr and about the intermediate states that have played
a significant role at certain near-resonant frequencies. Such information is
essential input for the screening of various theoretical models. Particularly
in Sr and a few other atoms, we have the opportunity to study the participation of bound doubly excited states as intermediate states, as well as
the role of intercombination transitions. For example, one of the angular
distributions reported by Feldmann and Welge [3] shows strong signs of
participation of a (two-photon excited) triplet intermediate state to three-
photon ionization, although the initial state is a singlet.

   The theoretical framework development in the preceding pages leads to
a number of predictions other than those already discussed. One example,
related to the modification of configuration interaction of the autoionizing
state by the field, could be called modification of configuration mixing of the
bound states by the field. Since many of the states of an atom like Sr are
not of a pure configuration, when they appear as (especially near-resonant)
intermediate states in a multiphoton transition, one must account for the
possible effect of the field on the mixing of the configurations. This would
arise because of virtual transitions coupling the particular intermediate state
to other (intermediate) states, very much like $\mathscr{E}^2 \mu_{cb}^{(2)}$, which modifies V_{cb}.
What the net result will be depends on relative signs and magnitudes of
the respective matrix elements. One of the possibilities, however, can be
the decoupling of configurations. In other words, under the influence of the

field, the particular intermediate state may participate as if the configurations were not mixed or were mixed by different amounts. If under such conditions one extracts configuration interaction parameters from, let us say, ionization of this state, the parameters would not necessarily correspond to the radiation-free (bare) atom. This would occur independently of whether the final step is autoionization or simply ionization, as long as a state with configuration mixing served as a near-resonant intermediate state. Consequently, such effects may also be sought in more typical multiphoton ionization experiments on atoms such as alkaline earths. A more detailed discussion of this and similar aspect can be found in Refs. [21] and [22].

What we have identified in this article as modification of configuration mixing is related to what has been called photon catalysis by Lau and Rhodes. [24]. Their work has a different context and motivation, but is fundamentally related to the modification of intramolecular interactions by the laser field. More recent calculations on H_2 that are in different ways related to the theme of this chapter and to the paper by Lau and Rhodes has been reported in two papers by Dastidar and Lambropoulos [25]. It should be kept in mind that if the intensity exceeds significantly the value at which the modification of configuration interaction is important, saturation of the autoionizing transition may occur. Then we have the problem of ac-Stark splitting, which we discussed earlier and showed that it requires the complete time-dependent solution of the problem and not simply a transition probability per unit time. Of course this presupposes the saturation of a two- or three-photon transition, which, however, is within the realm of possibility when the intensity is in the region of 10^{10} W cm^{-2} or more.

In this chapter we have concentrated on multiphoton excitation of autoionizing resonances and problems related to strong field effects. The autoionizing resonances in question are owing to configuration interaction and similar intra-atomic or intramolecular processes which are independent of the field. In the recent literature other types of autoionizinglike resonances have also been discussed [26]. They are related to laser spectroscopy because they require the presence of laser radiation for the creation of the resonances. But the resonant behavior of the profile and its similarity with autoionization rests critically upon the presence of the field. It is the field that brings about the coupling between bound states and continua, thus causing the appearance of the asymmetric profiles. In fact, two separate fields are necessary. These types of effects are no less interesting, but it should be made clear that autoionization in the sense of configuration or related interaction is not present in those cases. It can be discussed and observed in a one-electron atom where configuration interaction is totally absent. And it has been discussed in the context of one-electron-like atoms such as Cs.

ACKNOWLEDGMENTS

This work was supported in part by the National Science Foundation through Grant No. PHY-81-00251 and in part by the Österreichische Fonds sur Förderung der Wissenschaftlichen Forschung.

REFERENCES

1. I. S. Aleksakhin, N. B. Delone, I. P. Zapesochny, and V. V. Suran, *Zh. Eksp. Teor. Fiz.* **76**, 887 (1979); *Sov. Phys. JETP (Engl. Trans.)* **49**, 447 (1979).
2. D. Feldmann, J. Krautwald, S. L. Chin, A. von Hellfeld, and K. H. Welge, *J. Phys. B* **15**, 1663 (1982).
3. D. Feldmann and K. H. Welge, *J. Phys. B* **15**, 1651 (1982).
4. F. Fabre, G. Petite, P. Agostini, and M. Clement, *J. Phys. B* **14**, L667 (1981).
5. A. L. Huillier, L. A. Lompre, G. Mainfray, and C. Manus, *Phys. Rev. Lett.* **48**, 1814 (1982).
6. J. J. Wynne and J. P. Hermann, *Opt. Lett.* **4**, 106 (1979); J. A. Armstrong, J. J. Wynne, and P. Esherick, *J. Opt. Soc. Am.* **69**, 211 (1979).
7. W. E. Cooke, T. F. Gallagher, S. A. Edelstein, and R. M. Hill, *Phys. Rev. Lett.* **40**, 178 (1978); W. E. Cooke and T. F. Gallagher, *Phys. Rev. Lett.* **41**, 1648 (1978).
8. U. Fano, *Phys. Rev.* **124**, 1866 (1961).
9. P. L. Altick, *Phys. Rev.* **169**, 21 (1968).
10. M. J. Seaton, *Comments At. Mol. Physics* **2**, 37 (1970).
11. C. M. Lee and K. T. Lu, *Phys. Rev. A* **8**, 1241 (1973).
12. U. Fano, *J. Opt. Soc. Am.* **65**, 979 (1975).
13. C. A. Nicolaides and D. R. Beck, *Int. J. Quantum Chem.* **14**, 457 (1978).
14. F. H. Mies, *Phys. Rev. A* **20**, 1773 (1979).
15. P. Lambropoulos and P. Zoller, *Phys. Rev. A* **24**, 379 (1981).
16. L. Armstrong, C. E. Theodosiou, and M. J. Wall, *Phys. Rev. A* **18**, 2538 (1978).
17. P. Lambropoulos, *Appl. Opt.* **19** 3926 (1980).
18. K. Rzazewski and J. H. Eberly, *Phys. Rev. Lett.* **47**, 408 (1981).
19. G. S. Agarwal, S. L. Haan, K. Burnett, and J. Cooper, *Phys. Rev. Lett.* **48**, 1164 (1982).
20. See, for example, related articles *in* "Multiphoton Processes" (J. H. Eberly and P. Lambropoulos, eds.), Wiley, New York, 1978.
21. Y. S. Kim and P. Lambropoulos, *Phys. Rev. A* (to be published.)
22. Y. S. Kim and P. Lambropoulos, *Phys. Rev. Lett.* **49**, 1698 (1982).
23. W. E. Cooke (private communication).
24. A. M. F. Lau and C. K. Rhodes, *Phys. Rev. A* **16**, 2392 (1977); A. M. F. Lau, *Phys. Rev. A* **25**, 363 (1982).
25. R. K. Dastidar and P. Lambropoulos, *Chem. Phys. Lett.* **93**, 273 (1982).
26. Yu. I. Heller, V. F. Lukinykh, A. K. Popov, and V. V. Slabko, *Phys. Lett.* **A82A**, 4 (1981).

9

Creation of Doubly Charged Strontium Ions

D. FELDMANN, H.-J. KRAUTWALD, AND K. H. WELGE

Fakultät für Physik
Universität Bielefeld
Bielefeld, Federal Republic of Germany

I. INTRODUCTION

After the first experiments by Aleksakhin *et al.* (1978, 1979) on double ionization by the radiation of a Nd-glass laser, the question arose whether double ionization is a direct process in which both electrons are simultaneously excited through two electron-excited states and emitted, or a stepwise process in which, in a first step, singly charged ions are created and ionized again within the same laser pulse.

To shed some more light on this problem we have measured the creation of doubly charged strontium ions Sr^{2+} by radiation from a Nd-YAG laser and its harmonics; in addition, we have applied tunable radiation in a wavelength region around 561 nm.

In single ionization processes (see Mainfray and Manus, 1980, and references therein) it has been observed that the number of photons necessary for ionization from energy conservation is equal to the slope of the ion yield against intensity curve on a log–log scale. But if one or more steps of the ionization process are in resonance with excited levels of the atom, the slope drastically changes in the vicinity of the resonant wavelength. Therefore the intensity dependence can indicate whether the ionization is resonant or not, but the experimental results discussed below will show that the slopes of Sr^+ and Sr^{2+} yields at fixed wavelengths cannot give an unambiguous answer to the problem of distinguishing the two mechanisms, whereas the results obtained with tunable radiation around 561 nm can easily be explained by assuming a stepwise process.

Before we start to describe the experiments and discuss their results, we want to make some general remarks on the process of double ionization of Sr atoms.

Double ionization can, in principle, proceed by two different mechanisms:

(1) Stepwise ionization:

$$Sr + mhv \rightarrow Sr^+ + e^- \tag{1a}$$

$$Sr^+ + nhv \rightarrow Sr^{2+} + e^- \tag{1b}$$

(2) Direct two-electron ionization:

$$Sr + lhv \rightarrow Sr^{2+} + 2e^- \tag{2}$$

(m, n, and l denote the number of photons necessary for the process.)

The number of photons $m + n$ absorbed in process (1) can be equal to the number l in process (2), or it can be larger if the electron emitted in step (1a) carries away kinetic energy. Table I shows the numbers in the Sr case for the photon energies we used.

If no resonance and saturation occur, the Sr^{2+} yield $N^{2+}(I)$ should show an intensity I dependence of

$$N^{2+}(I) \sim I^{(m+n)} \quad \text{case (1)} \quad \text{and} \quad N^{2+}(I) \sim I^l \quad \text{case (2)}$$

If step (1a) is saturated, the yield should only have the intensity dependence of the unsaturated step (1b): $N^{2+}(I) \sim I^n$.

If resonances occur these intensity dependences derived from perturbation theory are no longer valid (Mainfray and Manus, 1980).

II. EXPERIMENT

Figure 1 shows the experimental setup. An atomic beam is perpendicularly crossed by the focused output from the laser. The ions are extracted from the interaction region by an electric field of about 20 V cm^{-1}, analyzed in a quadrupole mass spectrometer and detected by a multiplier. In addition to the quadrupole field mass selection, a time-of-flight discrimination is applied

Table I

The Minimum Numbers of Photons for the Processes (1a), (1b), (1), and (2)
Defined in the Text and the Experimental Slopes

Wavelength (nm)	Number of photons (process)				Experimental slopes			
	m (1a)	n (1b)[a]	$m + n$ (1)	l (2)	k^+	k^+ [a]	k^{2+}	k^{2+} [b]
1064	5	10	15	15	5.6 ± 1	5.0 ± 0.2	8 ± 1.5	10.1 ± 0.3
532	3	5	8	8	3.5 ± 1		5.5 ± 0.5	
355	2	4	6	5	2.3 ± 0.3		—	
266	2	3	5	4	1.6 ± 0.4		3.7 ± 1	

[a] The numbers n and $m + n$ are given for ionization via ground state Sr^+.
[b] Measured by Aleksakhin et al. (1979).

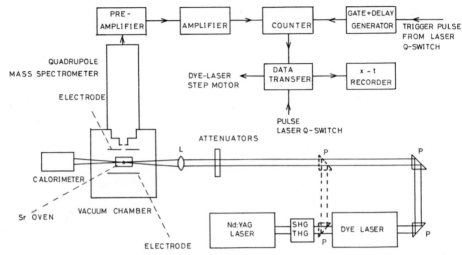

Fig. 1. Schematic experimental layout. [Reprinted with permission from Feldmann *et al.* (1982a). Copyright 1982 by The Institute of Physics.]

by using a gated counter, which only accepts pulses in a time interval delayed by the time of flight of the ions created on the electric potential of the interaction region and during the laser pulse.

The laser intensity can be linearly attenuated before the beam is focused. The average laser pulse energy for each intensity is measured by a calorimeter. The laser intensities are calculated from the experimentally determined pulse energy, pulse duration, beam divergence, and focal length of the lens. Table II gives typical experimental parameters.

III. RESULTS AND DISCUSSION

The ion signals are not corrected for detection efficiency, which may be different for Sr^+ and Sr^{2+}. We usually optimized the ion optics for Sr^{2+} detection and did not change it for Sr^+ detection.

For the fixed wavelengths, the ion yields versus laser intensities are shown in Figs. 2–5 on a log–log scale. From a fit of the linear parts of the curves, the slopes k_{exp} given in Table I have been derived. The bent part of curves is drawn by hand. Table I also includes the numbers m, n, l introduced above.

For a discussion of the double ionization with respect to a direct or stepwise process, we have to consider the single ionization too, because in the stepwise process (1), it is the first part (1a) and in a direct ionization (2), it would be a competing channel.

Table II

Typical Experimental Parameters[a]

	Wavelength (nm)				
	1064	532	355	266	∼561
Laser pulse energy (mJ)	≤150	≤25	≤10	≤1.5	∼10
Laser pulse duration (ns)	∼8–9	∼6–7	∼5–6	∼4–5	∼5–6
Laser bandwidth (cm⁻¹)	≤2				≤0.5
Estimated laser intensity in the focus of a 100-mm lens (GW cm⁻²)	≤130	≤25	≤12	≤2	∼40

[a] Number of Sr atoms in the focal volume ≈ 10^3.

A. Sr⁺ Creation

The single ionization of Sr at these wavelengths is essentially a non-resonant process, as is evident from the agreement between the number m of photons necessary and the experimental slopes k^+; this is corroborated by the lack of levels of SrI in resonance with multiples of the photon energies (Moore, 1958). For 1064 and 532 nm we observe a decreasing slope of the Sr⁺ yield at high intensities, which indicates saturation when a considerable part of the atoms within the interaction region is ionized. The "early" saturation at 532 nm may be due to an autoionizing 3-photon resonant level, which

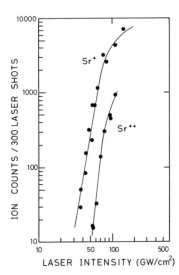

Fig. 2. Log–log plots of the ion counts versus laser intensity at $\lambda = 1064$ nm. [Reprinted with permission from Feldmann *et al.* (1982a). Copyright 1982 by The Institute of Physics.]

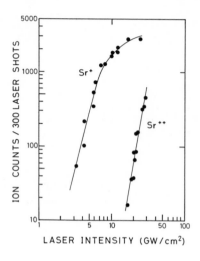

can come into resonance at high intensities by power shift and broadening (Feldmann *et al.*, 1982a). From this saturation behavior we can derive values of the generalized cross section for multiphoton ionization of step (1a):

$$\hat{\sigma}_3(532) = 10^{-78} \quad \text{cm}^6 \, \text{s}^2$$

$$\hat{\sigma}_5(1064) = 10^{-141.5} \quad \text{cm}^{10} \, \text{s}^4$$

These numbers should only be considered as a rough estimate of the exponents. The first value may be compared with a first rough estimate by Lambropoulos (1982, private communication) of $\hat{\sigma}_3 (532) \approx 10^{-80}$ cm^6 s^2

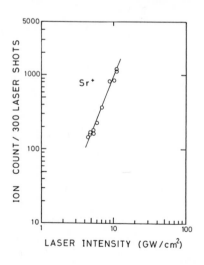

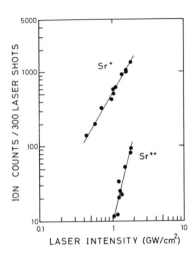

Fig. 5. Log–log plots of the ion counts versus laser intensity at $\lambda = 266$ nm. [Reprinted with permission from Feldmann *et al.* (1982a). Copyright 1982 by The Institute of Physics.]

and the second one with a determination of Aleksakhin *et al.* (1979) at the Nd-glass laser wavelength: $\hat{\sigma}_5 = 10^{-140.6^{+1.8}_{-1.7}}$.

In a recent experiment L'Huillier *et al.* (1982) have observed multiple ionization of Kr at intensities between 10^{13} and 10^{14} W/cm^2. Above saturation they find intensity dependences proportional to $I^{3/2}$ and I^3 which can be attributed to an increasing focal volume. Our experiments at 532 nm show a much smaller slope at the maximum intensities, which may justify the derivation of an estimated cross section from this saturation curve. One main reason for this different experimental saturation behavior seems to be the different ion extraction optics used in the two experiments. Ours has a high local resolution because it is a modified version of a SIMS–optic combined with a quadrupole mass filter so it only accepts ions from the central part of the focus whereas for the Kr experiments a comparatively high extraction field has been applied for ion extraction in combination with a TOF spectrometer. In principle this latter set-up can accept ions from a larger volume.

B. Sr²⁺ Creation

Some general findings for all wavelengths at which Sr^{2+} has been observed can be seen in Figs. 2, 3, and 5. At the intensities of our experiments, the Sr^{2+} yield is one or two orders of magnitude lower than the Sr^+ yield. At 355 nm, no Sr^{2+} ions could be detected even when the Sr density was raised by a factor of 10.

Comparing the experimental slopes k^{2+} with the numbers l of photons necessary from energy considerations, one can see that for 1064 and 532 nm k^{2+} is definitely lower. This indicates that resonances are involved. As discussed above, such resonances do not occur below the first ionization limit, therefore they have to lie between the first and second ionization limit; but since their character is still open, they can be two-electron excited states of the atom (autoionizing states) or levels of SrII.

In the following some arguments will be discussed that might help to distinguish between stepwise or direct ionization.

(a) It has already been mentioned by Aleksakhin *et al.* (1978) that the Sr^{2+} creation should eventually influence the Sr^+ yield by lowering its slope. Without detailed theory, this effect cannot give a decision among the two mechanisms: for process (1) the number of Sr^+ ions should be reduced when some of them are ionized, but also for direct ionization (2) the Sr^{2+} creation is a competitive channel, which reduces the number of atoms available for single ionization.

In our experimental results the effect cannot be found primarily because the Sr^{2+} yield is at least one order of magnitude lower than the Sr^+ yield for the laser intensities we could use.

(b) For 1064 and 532 nm we observe saturation at high intensities. The difference at the two wavelengths is that for 1064 nm, the Sr^+ saturation is paralleled by a decreasing slope of the Sr^{2+} yield, whereas for 532 nm, a measurable number of Sr^{2+} ions is detected only at high intensities where the Sr^+ yield curve is rather flat.

The behavior at 1064 nm can be explained by both mechanisms: The direct creation of Sr^{2+} should become less efficient if most atoms are singly ionized and are no longer available for direct double ionization. On the other hand, when in a stepwise process the first step becomes less efficient also the second part of the process (1b) should no longer rise as steeply as it did for low intensities.

(c) For 532 nm one can argue in favor of a stepwise process: Sr^{2+} is created only when the Sr^+ yield saturates; in this case only the number of photons necessary for the second step (1b) should determine the slope of the Sr^{2+} yield, so one should compare the number $n = 5$ (third column of Table I) with the experimental slope $k^{2+} = (5.5 \pm 0.5)$ (column 8 of Table I). They agree within the experimental uncertainty.

But this does not strictly exclude the possibility of a resonant direct process.

The low Sr^{2+} yield at 355 nm (we could not detect Sr^{2+} ions even at a ten times higher particle density) only indicates that resonances seem to help at the other wavelengths.

The experimental slope for Sr^{2+} at 266 nm is uncertain by ± 1 and no saturation is observed. The value of 3.7 ± 1 is closer to $l = 4$ for a direct process but does not exclude a possible stepwise process.

This discussion shows that the slopes of the Sr^{2+} curves do not allow one to decide between the two models. This was the reason for an experiment with tunable radiation from a dye laser. Two wavelength regions have been investigated.

(a) In the vicinity of the second harmonic of the YAG laser around 530 nm, we could only reach intensities up to 5 GW cm^{-2}, which was insufficient to produce detectable amounts of Sr^{2+}. These experiments, described in more detail by Feldmann et al. (1982a), showed that the single ionization of Sr at 532 nm is assisted by a nearby three-photon resonance at high intensities, which seems to be the reason for the high Sr^+ yield and early saturation of the Sr^+ curve at 532 nm (Fig. 3).

(b) Around 561 nm, our dye laser has its maximum output and we have measured the ionization spectra (the wavelength dependence of the ion yield) as shown in Fig. 6.

For the intensity of 40 GW cm^{-2} the number of Sr^+ ions is more than 20 times higher than the number of Sr^{2+} ions. The Sr^{2+} curve shows two broad maxima in this wavelength region, which are not simply correlated with maxima of the Sr^+ curve.

The following explanation can be given for these maxima: Fig. 7 shows an energy level diagram of SrI and SrII including only levels relevant for this discussion. The figure also shows two ionization pathways for which the photons are indicated by arrows.

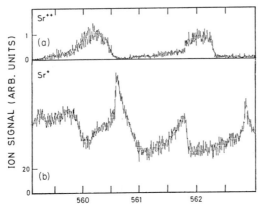

LASER WAVELENGTH (nm)

Fig. 6. Ionization spectrum of strontium atoms: (a) Sr^{2+} yield; (b) Sr^+ yield; laser intensity ≈ 40 GW cm^{-2}. (The vertical scale is different for the two curves.) [Reprinted with permission from Feldmann et al. (1982b). Copyright 1982 by The Institute of Physics.]

Three photons of 560-nm radiation can ionize Sr to give Sr^+ in its ground state $^2S_{1/2}$. Within the same laser pulse $Sr^+(^2S_{1/2})$ can be ionized by five more photons. The maximum in the Sr^{2+} yield is due to a resonance after four photons with the $4g\,^2G_{9/2}$ level of SrII (Moore, 1958). Power broadening and shift can be reasons for the width of the peak and its asymmetric shape, and it is an open question if the departing electron from the first ionization has a measurable influence on the position and form of the peak.

A resonant pathway for 562-nm photons is also shown in Fig. 7: one has to assume that four photons ionize Sr to give Sr^+ in its excited $5p\,^2P_{3/2}$ state. This requires the assumption that the fourth photon is absorbed in the ionization continuum. Such an assumption seems to be reasonable because the absorption of many excess photons above the ionization limit has recently been observed experimentally by Agostini *et al.* (1981) and Kruit *et al.* (1981). In those experiments the excess energy has been observed as kinetic energy of the departing electrons, whereas in our case we assume that excess energy is stored in the ion. From this $^2P_{3/2}$ state Sr^+ can be ionized again by four photons and the process is enhanced by the three-photon resonant state $9s\,^2S_{1/2}$ of SrII.

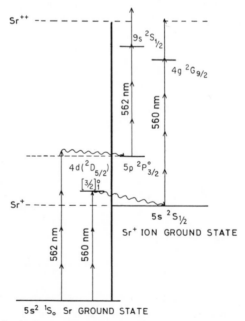

Fig. 7. Energy levels diagram of SrI and SrII (for details see text). [Reprinted with permission from Feldmann *et al.* (1982b). Copyright 1982 by The Institute of Physics.]

Reasons for the width of the peak can be the same as those mentioned above.

On its short wavelength shoulder the 562-nm peak shows a small steplike decrease at the wavelength of a resonance in the Sr^+ ionization spectrum below. From the electron energy spectrum we have measured with lower laser intensities (Feldmann and Welge, 1982), we have found that at this peak Sr^+ is preferably created in its ground state 5s ^2S and first excited 4d ^2D state. Ions in these states cannot be further ionized along the resonant path described above, which starts from the 5P $^2P_{3/2}$ state, so these favored ionization channels reduce the production of ions in the $^2P_{3/2}$ state thus causing the drop observed in the Sr^{2+} yield.

IV. CONCLUSION

Our experimental results show that the intensity dependences of Sr^{2+} creation at fixed photon energies give strong evidence that resonances are important for double ionization. For Ba atoms such a resonant character of double ionization by Nd-glass laser radiation has recently been observed by Delone et al. (1982). From our results we cannot decide if direct ionization or stepwise ionization occurs at these fixed wavelengths, but our ionization spectra obtained with tunable radiation at around 561 nm can easily be interpreted if stepwise ionization is assumed with resonances in the second ionization step.

REFERENCES

Agostini, P., Clement, M., Fabre, F., and Petite, G. (1981). J. Phys. B 14, L491.
Aleksakhin, I. S., Zapesochnyi, I. P., and Suran, V. V. (1978). JETP Lett. (Engl. Transl.) 26, 11.
Aleksakhin, I. S., Delone, N. B., Zapeschnyi, I. P., and Suran, V. V. (1979). Sov. Phys. JETP (Engl. Transl.) 49, 447.
Delone, N. B., Bondar, I. I., Suran, V. V., and Zon, B. A. (1982). Opt. Commun. 40, 268.
Feldmann, D., and Welge, K. H. (1982). J. Phys. B 15, 1651.
Feldmann, D., Krautwald, J., Chin, S. L., von Hellfeld, A., and Welge, K. H. (1982a). J. Phys. B 15, 1663.
Feldmann, D., Krautwald, J., and Welge, K. H. (1982b). J. Phys. B 15, L529.
Kruit, P., Kimman, J., and Van der Wiel, M. J. (1981). J. Phys. B 14, L597.
L'Huillier, A., Lompre, L. A., Mainfray, G., and Manus, C. (1982). Phys. Rev. Lett. 48, 1814.
Mainfray, G., and Manus, C. (1980). Appl. Opt. 19, 3934 and references therein.
Moore, C. E. (1958). "Atomic Energy Levels," NBS Circular 467. Natl. Bur. Stand., Washington, D. C.

10

Many-Electron Processes in Nonlinear Ionization of Atoms

N. B. DELONE

General Physics Institute
Academy of Sciences of the USSR
Moscow, USSR

V. V. SURAN

Uzhgorod State University
Uzhgorod, USSR

B. A. ZON

Department of Physics
Voronezh State University
Voronezh, USSR

I. INTRODUCTION

Since the process of multiphoton atomic ionization was revealed [1] and to the present, alkaline and noble gas atoms have been the object of investigation in the majority of the experiments. The interest in the latter atoms was explained by the relative simplicity of the use of a target in the form of a rarified gas; alkaline atoms were of interest because of their relatively low ionization potentials [1] and, accordingly, the relatively low field strength at which ionization could be observed.

In the experiments with alkaline and noble gas atoms, the question of the possibility of doubly charged ion production never arose. Indeed, in the case of alkaline atoms, second ionization potentials are almost an order of magnitude larger than first ones and range from 20–50 eV; second ionization potentials of noble gases have the same value. This means that within a small range of frequencies, the detachment of a second electron, and especially of two electrons simultaneously, requires the absorption of a large number of photons. This process can take place only in extremely strong fields, much stronger than those in which the detachment of a first electron is observed. A theoretical analysis of the possibilities of doubly charged ion creation under multiphoton ionization of atoms has not been carried out until recently. In ionization of noble gas atoms, the many-electron structure of the outer shell could also manifest itself in the multiphoton cross section of one-electron ionization. In most cases, however, one could not draw any definite conclusions concerning the role of many-electron structure of noble gas atoms because of an insufficient experimental accuracy of direct multiphoton noble gas ionization cross-section measurements.

Thus although much interesting information on the multiphoton ionization process was obtained both from experiments with alkaline atoms and from those with noble gas atoms [2, 3, 4], all this information referred to the detachment of one outer electron. The question of the role of many-electron atomic structure in multiphoton processes remained open.

The situation changed radically as soon as research workers turned to the investigation of the multiphoton ionization of alkaline-earth atoms and lanthanides. In the first experiments conducted by the group of Zapesochny (Uzhgorod University, USSR) on multiphoton ionization of strontium [5], barium [6], and samarium [7] atoms, each singly charged and doubly charged ions of the above-mentioned elements were registered. All these atoms are distinguished by relatively low first (5.2–6.5 eV) and second ($\simeq 10$ eV) ionization potentials. Ionization of these atoms was observed in the radiation field of Nd-glass laser at a frequency of 9450 ± 7 cm^{-1} and a field strength of $\sim 10^7$ V cm^{-1}. Ions were produced in the atomic-beam–laser-radiation-beam intersection zone and were analyzed by a time-of-flight mass spectrometer. The experiments were conducted under the following three conditions:

(1) a low (~ 1 eV) kinetic energy (E_k) of the electrons produced in the multiphoton ionization of atoms ($E_k = K_0 \hbar\omega - I \lesssim 1$ eV).

(2) an order of magnitude lower kinetic energy acquired by electrons in ponderomotive acceleration in the region of laser radiation focusing.

(3) a small atomic beam density of the order of 10^{10}–10^{12} atoms cm^{-3}.

These three factors altogether excluded any secondary effects that could arise after the action of laser radiation pulse. Therefore it was clear that doubly charged ions are formed directly in the process of laser radiation interaction with neutral atoms. A short laser pulse ~ 10 ns and a small beam intersection zone (~ 1 mm) clearly indicated that no secondary effects can take place during the action of the radiation; i.e., doubly charged ions are created by the interaction of radiation with an isolated atom.

Values of the first and second ionization potentials of the above-mentioned atoms, as well as the frequency and strength of the radiation field at which doubly charged A^{2+} ion formation was observed, show that ionization occurs under conditions when the adiabatic parameter $\gamma \gg 1$, i.e, the ionization process is of multiphoton character [2].

The doubly charged ion output amplitude observed in these experiments was relatively large—10^{-1}–10^{-3} of the singly charged ion output amplitude. This made it possible to investigate the physical nature of this new phenomenon.

The discovery of doubly charged ions in these first experiments [5–7] gave rise to a new trend in the studies of the process of multiphoton ionization of atoms—a trend connected with the study of many-electron effects. The well-known specific features of the spectra of alkaline-earth atoms and lanthanides, namely, the presence in the excited state spectra of two-electron bound and autoionization states, pointed to their qualitative difference from the spectra of alkaline atoms. The presence of several electrons in the outer shell could also play an essential role.

Before presenting and discussing the experimental data and theoretical conclusions that refer to the process of nonlinear many-electron ionization of atoms and, in particular, to the process of doubly charged ion production, it is necessary to briefly examine the main information concerning the spectra of excited states of these atoms, as well as many-electron effects in photoionization (one-photon ionization) of many-electron atoms. These data will be necessary in the analysis of experimental data on multiphoton processes in many-electron atoms.

II. SPECIFIC FEATURES OF THE SPECTRA OF
ALKALINE-EARTH ATOMS

It is well known that in the process of multiphoton ionization of atoms, an essential role is played by the spectrum of excited electron states. When ionization is of a direct character, the spectrum of excited states determines the multiphoton cross section for the bound–free electron transition. The resonance ionization occurs when there appears an intermediate resonance with a certain excited electron state. Therefore, analysis of any experimental data on multiphoton ionization of atoms is based on the information about the atomic spectrum. One should keep in mind that under the action of the external field, not only the transition of an electron from one state to another takes place, but the transition energy may be varied by an atomic level shift and splitting in the field. In strong fields the perturbation of electron states is large, so that for the analysis of experimental data one cannot use tabulated data on the nonperturbed atomic spectrum.

Now let us turn to the data on specific features of the spectra of alkaline-earth spectra and their perturbation in the radiation field.

A. Two-Electron Bound States

The alkaline-earth atomic spectra essentially differ from those of alkaline and noble gas atoms. The alkaline-earth atoms exhibit bound states owing to excitation of either one or two outer s-electrons (the so-called irregular terms) [8]. The states due to two-electron excitation lie, as a rule, above the potential of single ionization of the atom, whereas in alkaline-earth elements part of these states is located below the potential of single ionization. The data on two-electron bound states of magnesium, calcium, strontium, and barium atoms are tabulated in Ref. [9] and are also presented in Refs. [10, 11], in which both highly excited one-electron states and irregular terms were investigated by the method of multiphoton excitations through intermediate states. In Table I electron configurations of two-electron states, their number and order of appearance are listed.

As follows from the data listed in Table I, as the atomic number of the element increases, the number of irregular terms enlarges, the atomic spectrum becomes more complicated. Note that the limits of the series of the irregular terms presented in Table I are the lowest ion states, specifically, the metastable n^2D and the resonant n^2P states. As far as other ionic states of these atoms are concerned, they are the limits of the series of autoionization states (see Section II.B).

The widths and, accordingly, the lifetimes of two-electron states are approximately the same as in one-electron states.

Table I

Data on the Spectra of Bound Two-Electron States

Mg		Ca		Sr		Ba	
Electron configuration	Number	Electron configuration	Number	Electron configuration	Number	Electron configuration	Number
$3p^2$	1	$3d4p$	5	$4d5p$	5	$5d6p$	6
		$4p^2$	3	$5p^2$	3	$5d^2$	2
		$3d5s$	1	$4d^2$	1	$5d7s$	2
		$3d^2$	1			$6p^2$	3
						$5d6d$	9
						$5d7p$	6
						$5d4f$	3
						$5d8p$	2

The presence of bound two-electron states in the atomic spectrum shows that it is possible, in principle, for two electrons simultaneously to acquire energy up to the value corresponding to the first ionization potential. The probability of atom excitation to a two-electron bound state is rather high. For example, the probabilities of photoexcitations of two-electron and one-electron states are identical in the order of magnitude. One should assume, therefore, that two-electron bound states can also play an essential role in multiphoton excitation or in multiphoton ionization of the atom.

B. Autoionization States

Another characteristic feature of the alkaline-earth atomic spectra is the presence of discrete states, which are located above the first ionization potential and are due to excitation of two outer s electrons. Since above the atom ionization boundary there exists discrete and continuous spectra, there is a certain probability for the atom to pass over from a discrete spectrum to a continuum, i.e., to become ionized. This phenomenon is referred to as autoionization.

If one electron is in an excited state $n_1 l_1$ and the second is excited to higher and higher states $n_2 l_2$, the series of excited terms will converge to the boundary whose value is equal to the sum of the energy of ionization of the atom and the excitation energy of the ion in the state $n_1 l_1$. Thus in alkaline-earth atoms, all the levels of singly charged ions are limits of the series of two-electron excitation of outer s electrons. Accordingly, in the energy range from first to second ionization potential, along with ion states, there exists a large number of series of autoionization states. Note that in all the cases the principal quantum number and the orbital number of one electron from a series of autoionization states coincide with corresponding numbers of the ion state, which is the limit in this series.

All the experimental data obtained before 1980 on the absorption spectra of alkaline-earth atoms in the vacuum UV range are presented in detail in Ref. [12]. Autoionization states were also investigated by the methods of multiphoton absorption through intermediate real [13–15] and virtual [16] states. Note that the multiphoton spectroscopy method (as distinguished from one-photon spectroscopy) makes it possible to investigate states whose parity coincides with that of the ground state and also states with a large total angular momenta.

In spite of the existence of different methods for investigating autoionization states, only those have as of yet been practically observed in alkaline-earth atoms; the limits of the series of which are low-lying $n\,^2D$ and partially $n\,^2P$ states of ions [17, 18], i.e., states near the first ionization potential. As to autoionization states due to excitation of two outer s electrons located above the ion state $n\,^2P$, no experimental data are currently available.

Table II

Autoionization States of the Ba Atom
$5dnp\ ^1P_1^0$ *and the Radiation Frequency* ω
at Which the Five-Photon
Excitation is Possible

n	$E_n(\text{cm}^{-1})$	$\omega(\text{cm}^{-1})$
15	46854	9371
16	46987	9397
17	47090	9418
18	47173	9434
19	47240	9448
20	47296	9459
21	47342	9468
22	47381	9476
23	47414	9483
24	47442	9488
25	47466	9493
26	47487	9499

Autoionization states of a barium atom of the series $5dnp$ were studied in Ref. [19] by the one-photon absorption method. Table II presents those states of the series $5dnp\ ^1P_0^1$ with which a five-photon resonance of neodymium laser radiation may occur. We shall need these data below for the analysis of experiments (see Section V).

The photoabsorption cross section of autoionization states of calcium, strontium, and barium atoms exceeds, in all the cases, the photoionization continuum absorption cross section approximately by an order of magnitude [17, 18].

The experimental data on the widths Γ of some autoionization states are listed in Table III. Some conclusions can be drawn from these results. First, the autoionization state widths exceed bound state widths ($\Gamma \simeq 10^{-3}\ \text{cm}^{-1}$). Second, states of the calcium atom have the largest widths, and as the atomic number of the element increases, their widths decrease. Note that the authors of Ref. [23] show, proceeding from model considerations, that $\Gamma \sim Z^{-2}$, where Z is the atomic number of an element. Third, the states following the atom ionization threshold have larger widths; as the state energy increases, their widths decrease.

The authors of Ref. [13] investigate the dependences of the widths of the autoionization states of the strontium atom on the effective principal quantum number n^* and the orbital momentum of one of the outer electrons. The state widths $\Gamma \sim (n^*)^{-3}$. As the orbital momentum increases, the width Γ also decreases sharply.

Table III

Experimental Data on the Widths Γ of Autoionization States

Ca^a		Sr^b		Ba^c	
State	$\Gamma(\text{cm}^{-1})$	State	$\Gamma(\text{cm}^{-1})$	State	$\Gamma(\text{cm}^{-1})$
$3d5p\,^1P_1^0$	615	$4d6p\,^1P_1^0$	542	$5d8p\,^1P_1^0$	165
$3d6p\,^1P_1^0$	490	$4d7p\,^1P_1^0$	193	$5d9p\,^1P_1^0$	43
$3d7p\,^1P_1^0$	275	$4d8p\,^1P_1^0$	168	$5d10p\,^1P_1^0$	23
$3d8p\,^1P_1^0$	160	$4d9p\,^1P_1^0$	57	$5d11p\,^1P_1^0$	12
$3d9p\,^1P_1^0$	100				
$3d10p\,^1P_1^0$	80				
$3d11p\,^1P_1^0$	40				

[a] Reference [20].
[b] Reference [21].
[c] Reference [22].

Only a small part of the autoionization states of the atoms in question has been experimentally studied and identified at the present (see Ref. [12]). The most important thing is that the experimental data available do not cover the whole range from the first to the second ionization potential. Therefore to have an insight into the general picture of the autoionization state spectrum, we turn to the results of theoretical calculations. Unfortunately, it is only one calculation of the spectrum of autoionization states of the barium atom [24] that is of interest to us. Modern methods permit calculations of the autoionization state spectra with a sufficiently high relative accuracy. In this case, calculations were carried out by the Hartree–Fock–Dirac method in the frozen atomic core approximation with an account taken of relativistic effects, which essentially affect the values of quantum defects of excited electrons. In spite of a high relative accuracy of the calculations, their absolute accuracy is insufficient for direct comparison of computational data with the results of experiments (see Table IV). On the whole, however, the calculation gives the general picture of the spectrum—in the energy range from the first to the second ionization potential of the barium atom, the autoionization state density proves to be rather high. In particular, in the vicinity of all the energies $K\hbar\omega$, where ω is the neodymium laser radiation frequency, there exist a considerable number of autoionization states.

Of all the above-mentioned data on the autoionization state spectra, the most important data concerning the indicated effect of nonlinear ionization of many-electron atoms are those on a high density of autoionization states in the energy range between the first and second ionization potentials. These

Table IV

*Comparison of Computational [24] and Experimental [19] Data
on the Spectrum of Autoionization States of the Ba Atom*

Notation	I	I_{at}	E_{theor} (cm^{-1})	E_{exp} (cm^{-1})	ΔE
$5p^6 5d_{5/2} np_{3/2}$					
$5p^6 5d_{5/2} 7p_{3/2}$	2.5	1	38794.733	38499.852	294.88
$5p^6 5d_{5/2} 8p_{3/2}$	—	1	42322.00	42012	310
$5p^6 5d_{5/2} 9p_{3/2}$	—	1	44100.99	43911	190
$5p^6 5d_{3/2} np_{3/2}$					
$5p^6 5d_{3/2} 7p_{3/2}$	1.5	1	37421.985	36990.016	431.97
$5p^6 5d_{3/2} 8p_{3/2}$	—	—	41265.13	41096.8	168.3
$5p^6 5d_{3/2} 9P_{3/2}$	—	—	43162.57	43027	135.16

data, along with those on two-electron bound states, show that under the influence of the external field, two outer electrons can make virtual or real transitions along the two-electron state spectrum up to the second ionization potential, when both the electrons pass over to a continuum and a doubly charged ion is formed. Relaxation of autoionization states in the case of their real occupation is a competing process leading to singly charged ion formation.

The above analysis of the data on the spectra of alkaline-earth atoms shows that two-electron bound and autoionization states can manifest themselves essentially in the nonlinear ionization process. In our opinion, this is what essentially makes the process of multiphoton ionization of these atoms different from alkaline and noble gas atoms.

C. Perturbation of the Atomic Spectrum by the External Field

The phenomenon of atomic spectrum perturbation by an external variable field is rather well known [2, 3]. This phenomenon by itself provides valuable information concerning the interaction between atom and field. At the same time it determines the character of the resonant ionization process and thus hampers the analysis of experimental data on multiphoton processes. The difficulty is that at large laser radiation field strengths, a strong shift and the splitting of atomic levels may occur. Accordingly, the spectrum of excited electron states of a nonperturbed atom turns out to be of little use for interpretation of the experimental results.

The shift of the level n in the field in a first approximation is determined by dynamical polarizability of this level $\alpha_n(\omega)$:

$$E_n = -\tfrac{1}{4}\alpha_n(\omega)\mathscr{E}^2$$

Here for simplicity we assume the level n to possess a zero total momentum, thus it does not split in the field. Besides, it is assumed that there is no resonant mixing of the level n with other atomic states. These, as well as some other more general cases of atomic spectrum perturbation in a variable field, are considered in detail in [2].

As $\omega \to 0$, the quantity $\alpha_n(\omega) \to \alpha_n(0)$, i.e., tends to static polarizability. At $\omega \gtrsim E_n$ the electron may practically be considered free, $\alpha_n(\omega) \to -Z/\omega^2$, i.e., tends to the oscillation energy of a free electron in the wave field. At intermediate values of ω, the quantity α depends essentially on ω, increases as resonances appear, and changes the sign in the resonances. Hence the quantity $\alpha(\omega)$ can be calculated only when numerical methods are used. In obtaining experimental values of $\alpha_n(\omega)$, particularly for excited atomic states, we encounter considerable difficulties. Only some single measurements of the kind performed on alkaline atoms are known at the present time. As concerns theoretical calculations of the quantities $\alpha_n(\omega)$, the proposed computational methods are limited, as a rule, either to the ground states of the atoms or to atoms with one valence electron. In particular, we do not know any calculations of $\alpha_n(\omega)$ for alkaline-earth atoms and lanthanides.

For this reason, for alkaline-earth atoms we have to restrict ourselves to the estimates of the quantity $\alpha_n(\omega)$. Such estimates are easily obtained for the ground states of atoms, as well as for highly excited states. For frequencies essentially lower than that of neodymium laser radiation, one can assume with a sufficient accuracy that the dynamical polarizability of the ground state $\alpha_n(\omega)$ coincides with the static value $\alpha_n(0)$. In turn, the quantities $\alpha_n(0)$ are known both from calculations and from direct experimental measurements. Table V presents the values of $\alpha_n(0)$ and the shifts of the ground states of alkaline-earth atoms in a direct field of strength $\sim 10^7$ V cm. As it is seen, the shifts are quite substantial. From general considerations it is clear that the shifts of excited atomic states will be still larger. Rather true estimates can also be obtained for highly excited states whose binding energy $E_n \lesssim \omega$. It follows from this relation that the field strength $\sim 10^7$ V cm^{-1}

Table V

Experimental Data on Static Polarizability of the Ground States of Alkaline Earth Atoms[a]

Atom	$\alpha_0(0)$ (a.u.)	Reference	ΔE (cm^{-1})
Mg	33	[25]	7
Ca	135	[26]	27
Ba	266 ± 21	[27]	54

[a] The shift of the ground state is calculated for the field strength 10^7 V cm^{-1}.

and the neodymium laser radiation frequency $\Delta E_n \simeq 10^2$ cm^{-1}. Therefore, the attempt to explain the experimentally observed resonances at the field strength $\sim 10^7$ V cm^{-1}, using the information on the spectrum of a nonperturbed atom, should be considered ungrounded.

Note that in the investigation of the process of multiphoton ionization of alkaline atoms, the fields of $\sim 10^6$ V cm^{-1} were used, in which the atomic level shifts proved to be essentially smaller. Although the shifts of the levels in this case exceed their natural width by several orders of magnitude, they (the shifts) are of the same order of magnitude as other factors determining the width of multiphoton resonances, for example, as the laser radiation spectrum width. For this reason, interpretation of resonances in multiphoton ionization of alkaline atoms face no difficulties. A similar situation arises in the investigation of alkaline-earth atoms and lanthanides when the second harmonic of an Nd laser is used since ionization is observed at the field strength $\sim 10^6$ V cm^{-1}. It should be noted, however, that in some cases the replacement of dynamic polarizability of ground states by their static polarizability may give underestimated values because the frequency of the second harmonic is closed to the first resonant frequency in the atomic spectrum.

Summarizing, one can state that only in relatively weak external fields $\sim 10^6$ V cm^{-1} can the tabulated data on nonperturbed atomic levels be used for interpretation of experimentally observed resonances in the ion output. Such a possibility is absent for large field strengths.

The situation is much worse with the data on perturbation of the autoionization state spectrum by an external variable field of the light frequency range. Neither experimental nor theoretical data of the kind practically exist. An exception is a recent experiment [28] devoted to a direct three-photon excitation of some autoionization states of the Sr atom by laser radiation with the wavelength of 5560–5640 Å. As the radiation field strength increased in the range 2–4 $\times$ 10^6 V cm^{-1}, the resonances broadened in the Sr$^+$ ion output approximately by an order of magnitude. The results of that paper show that the data on autoionization states of a nonperturbed atom can be used only in relatively weak fields. At the present time it is difficult to say to what extent the limit on the field strength (about 10^6 V cm^{-1}), which follows from Ref. [28], is universal. However, it should be noted that this limit is in a rather good agreement with the data for the above-mentioned bound states.

III. MANY-ELECTRON PHOTOIONIZATION OF ATOMS

The fact of multicharged ion formation in photoionization of atoms was known long ago. However, only one multicharged ion formation mechanism has been investigated for many years, namely the Auger process. This process is of an explicitly cascade character. At the first stage of the cascade, a hard photon takes away one atomic electron from the inner shell. The hole thus formed can be filled with electrons from outer shells, and the excess of energy at this second stage of the cascade can be lost both for photons, by radiative decay of atoms, and for other atomic electrons as a result of the Auger effect. This latter mechanism of the decay of atomic vacancy is just what leads to multicharged ion formation. The cascade character of the Auger process manifests itself in the discrete spectrum of Auger electrons.

Quite recently, the photoionization process leading to the formation of doubly charged [29] and triply charged [30] ions as a result of a single process of detachment of several electrons from the atom was discovered experimentally. In this case, the photon energy is continuously distributed among the ejected electrons. The process of a simultaneous detachment of two electrons manifests itself clearly in the formation of He^{2+} ions when neutral helium is radiated by hard quanta. Since the He atom contains only two electrons, the Auger effect in the He^+ ion is impossible.

It is obvious that the possibility to take away from the atom two or three electrons simultaneously is connected with interelectron correlation in atomic wave functions and is a specific many-body effect. Depending on the incident photon energy and energy distribution among the ejected electrons, it is either correlations only in the initial state or correlations both in the initial and the final states that may appear to be of importance.

The case when one of the electrons possesses a much higher energy than the rest of the electrons is the most simple for theoretical description. The process of detachment of slow electrons from the atoms may be associated with a very rapid change of a self-consistent atomic field, which took place after the detachment of a fast electron. In this limit the process is of the character of atom shake-off [31, 32] and is similar to that at alpha or beta decay of the nucleus. It is apparent that in this case interelectron correlations in the initial state are inessential, at least for doubly charged ion formation. It is also clear that in this limiting case the ratio of the cross section of the formation of doubly charged σ^{2+} ions to that of singly charged σ^+ ions is no longer dependent on the incident photon energy E_γ, when $E_\gamma \to \infty$. For the He atom, the ratio σ^{2+}/σ^+ in this limit makes up the value $\simeq 3-5\%$, for Ne and Ar, $\simeq 10-20\%$ [33].

In the case when the photon energy is distributed among the ejected electrons more or less equally, the theory of the phenomenon turns out to be rather complicated and the cross section of the process depends nonmonotonically on the photon energy. Without presenting here in detail the available theoretical and experimental results, we shall only point out that the cross section of the formation of doubly and triply charged ions of noble gases is of the order of $\sim 10^{-18}$ cm^2 [29, 30, 33].

As for the above-mentioned alkaline-earth atoms, the experimental data here are poorer and, what is even more important, they have not been used in so much detail as in the case of the helium atom. We shall only point out the case of photoionization of the barium atom, for which one observed an extremely large Ba^{2+} ion output exceeding the Ba^+ ion output at an ionizing radiation wavelength of about 580 Å [34].

Critical discussion of the present state of the problem of multiple ion production in photoionization of atoms is carried out in Ref. [35].

Specific nature of the process of photoionization (one-photon ionization) of atoms shows that the spectrum of excited states of an electron in the atom does not play a determining role in ionization. In photoionization it is not interelectron correlations in intermediate states (where they result, in particular, in two-electron states) that are essential, but those in the initial and in the final states. Analogous interelectron correlations may, in principle, also be observed in multiphoton ionization of atoms. However, in multiphoton ionization one should expect that the main effect must be connected with the existence of two-electron bound and autoionization states.

IV. FORMATION OF DOUBLY CHARGED IONS IN NONLINEAR IONIZATION OF MANY-ELECTRON ATOMS

Turning to the experimental data on the formation of doubly charged ions, it is relevant first to consider the results of experiments at a fixed laser radiation frequency. The point is that it is only under such conditions that the data for some many-electron atoms were obtained [in the above-cited papers [5] (Sr), [6] (Ba), [7] (Sm), and also in Refs. [36] (Ca) and [37] (Eu, Mg, Pb) at a frequency of 9450 cm^{-1}]. These data are summarized in Ref. [37]. Besides, Ref. [38] presents the results for the Sr atom obtained at a frequency of 9395 cm^{-1}.

The results published in [5–7, 36, 37] were obtained using radiation of a standard pulse neodymium-glass laser. The radiation frequency was 9450 ± 7 cm^{-1}; the radiation polarization was linear. The radiation field strength in the ion formation region was $6 \times 10^6 - 1 \times 10^7$ V cm^{-1}. For Ba, Sr, and Eu, the dependence of A^+ and A^{2+} ion output on the energy Q of the laser radiation pulse are measured (see Fig. 1). The laser working regime was such that the dependence $A(Q)$ was similar to that of the probability of ionization on the radiation intensity $\mathcal{W}(F)$ [39]. For Sm the Sm^{2+} ion formation only was registered. In the case of Ca, Mg, and Pb atoms the A^+ ions alone were observed. If A^{2+} ions were formed at all, their output was less than 10^{-3} of the A^+ ion output.

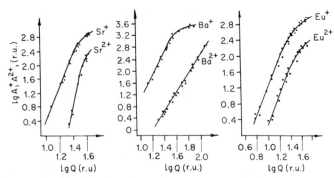

Fig. 1. Dependence of A^+ and A^{2+} ion output on the energy in the laser radiation pulse Q [37].

The data presented in [38] were obtained using linearly polarized radiation on YAG laser ($\omega = 9395 \pm 0.5 \text{ cm}^{-1}$). Sr^+ and Sr^{2+} ions were observed, the dependences $A(Q)$ were measured.

Thus first experiments [5–7, 36–38] on nonlinear ionization of a number of alkaline-earth atoms and lanthanides provided information about the formation of doubly charged ions, about the singly (A^+) to doubly (A^{2+}) charged ion output ratio, and about the dependence of the A^+, A^{2+} ion outputs on the radiation intensity.

First, we turn to the fact of the discovery of doubly charged ions. At the present time one cannot draw any conclusion from the fact that no doubly charged ions of the Ca, Mg, and Pb atoms were revealed. Strictly speaking, as has already been mentioned above, the experiment shows only that the A^+/A^{2+} ratio exceeded 10^3. The most probable reason for such a ratio for Ca and Mg is a strong resonance frequency dependence of the A^+ and A^{2+} ion output. The experiments with the Ba atom, to be described in Section V, have shown the existence of resonance dependence of Ba^+ and Ba^{2+} ion output and thus the existence of a strong frequency dependence of the ratio A^+/A^{2+}. It can be assumed that for these atoms at the radiation frequency of 9450 cm^{-1} there exist minima in A^{2+} ion output. Future experiments will show the validity of this statement. The specific features of electronic states of the Pb atom are that two-electron states are absent in the spectrum of this atom, while autoionization states are supposed to have very high energies. Therefore, doubly charged ion formation should not be expected in nonlinear ionization of the Pb atom.

Now let us turn to the data on the dependences $A(Q)$ (see Fig. 1). These dependences in log–log coordinates have a standard form, typical for the process of multiphoton ionization of atoms [39]. The change in the slope of curves in the region of large Q (i.e., in the region of strong fields) is connected with saturation in the ion output due to a high value of ionization probability $\mathscr{W}$, when $\mathscr{W} \geq 1/T$, where T is the laser pulse duration. It is well known that the slope of the curve $K = \delta \log A/\delta \log Q$, under conditions when $\mathscr{W} \ll 1/T$, characterizes the degree of nonlinearity of the direct ionization process (when no intermediate resonances appear) and is not directly connected with the degree of nonlinearity in the case of resonance ionization [39].

Therefore, in the absence of information concerning the frequency dependence of the ion output, one cannot draw any unambiguous conclusions from the values of K. The only information provided by the dependences $A(Q)$ is that the output of doubly charged ions is observed under conditions when there is no saturation in the output of singly charged ions.

The ratio A^+/A^{2+} measured in these experiments gives an unambiguous answer to the question concerning the character of the process of A^{2+} ion formation, i.e., whether this is a direct or a cascade process.

We refer to such a process as a direct one, when there exists a neutral atom in the initial state, and a doubly charged ion in the final state:

$$A + K_1\hbar\omega \rightarrow A^{2+} + 2e \tag{4.1}$$

If at first a singly charged ion is formed

$$A + K_2\hbar\omega \rightarrow A^+ + e \tag{4.2}$$

whose ionization causes the appearance of a doubly charged ion

$$A^+ + K_3\hbar\omega \rightarrow A^{2+} + e \tag{4.3}$$

We call such a process a cascade one. It should be noted that when speaking of a cascade process of A^{2+} ion formation, we mean that both A^+ and A^{2+} ions are produced in one laser pulse.

One should bear in mind that although the energy absorbed by an atom to produce an A^{2+} ion is identical in both cases and, accordingly, the numbers of absorbed photons ($K_1 = K_2 + K_3$) are the same, there are essential differences between the cascade and the direct processes. The most important difference is in the spectra of excited states, which determine the probabilities of corresponding multiphoton transitions. So in the cascade process the probability of A^{2+} ion formation is determined by one-electron spectra of excited states of a neutral atom [reaction (4.2)] and of a singly charged ion [reaction (4.3)]. In the case of a direct ionization process, the probability of A^{2+} ion formation is determined by the spectrum of two-electron states (bound and autoionization) of a neutral atom. The difference in the character of these spectra (different state density, binding energy, width, different perturbations of these states by the radiation field) is responsible not only for the quantitative difference in the probabilities of A^{2+} ion formation, but also for a qualitative difference in the nature of cascade and direct ionization processes.

As has already been said above, the frequency ω and the radiation field strength ε, as well as the values of the ionization potentials I for the reactions (4.1–4.3), show that the adiabatic parameter $\gamma = \omega\sqrt{I}/\varepsilon \gg 1$, i.e., nonlinear ionization occurs in the multiphoton case [2]. Therefore, to analyze the experimental data on the relation between ion outputs, one can use the well-known values of multiphoton cross sections.

We should recall that in ionization of alkaline-earth atoms and lanthanides in the field of a Nd: glass laser radiation ($\hbar\omega \simeq 1.2$ eV), the production of singly charged ions (4.2) is a five- or six-photon process, and the production of doubly charged ions owing to ionization of singly charged ions (4.3) is a ten- and more-photon process.

Analysis of the well-known data on multiphoton cross sections shows that the observed relation between the A^+ and A^{2+} ion outputs strongly contradicts the assumption concerning the cascade ionization processes (4.2) and (4.3). Suppose that both the transitions (4.2) and (4.3) are of direct character, i.e., no intermediate resonances occur. Then the probabilities $\mathscr{W}_K$ of these transitions are described by the known power law

$$\mathscr{W}_K = \alpha_K F^K \tag{4.4}$$

where α_K is a K-photon cross section, F is the radiation intensity, K is the number of absorbed photons. In the reaction (4.2) of A^+ ion formation, let the quantity $K = K_2 = 5$, and in the reaction (4.3) of A^{2+} ion formation, let the quantity $K = K_3 = 11$. (For these values of K there are sufficiently accurate data on multiphoton cross sections.) Let us find the probabilities $\mathscr{W}_5$ and $\mathscr{W}_{11}$ in the field $\mathscr{E} \sim 10^7$ V cm^{-1}, i.e., at the intensity $F \sim 10^{30}$ photon cm^{-2} s^{-1}. We shall use for the quantities α_5 and α_{11} the results of experiments and calculations. We assume $\alpha_5 = 10^{-140 \pm 1}$ cm^{10} s^4 [40, 41], and $\alpha_{11} = 10^{-336 \pm 2}$ cm^{22} s^{10} [42, 43]. It is easily seen from (4.4) that at the above-mentioned values of α_5, α_{11}, and $F = 10^{30}$ photon cm^{-2} s^{-1}, there follows the value of the ratio of the probabilities

$$\frac{\mathscr{W}_5}{\mathscr{W}_{11}} = \frac{10^{10}}{10^{-6}} = 10^{16} \tag{4.5}$$

Note that this ratio is inversely proportional to the radiation intensity.

The quantity (4.5) will determine the ratio A^+/A^{2+} of ion outputs under conditions when $\mathscr{W}_5 \geq 1/T$, the A^+ ion output is saturated. Only under these conditions will the A^+ ion density in the ionization volume be of the order of the density of neutral atoms. In the opposite case, when $\mathscr{W}_5 < 1/T$, the ratio A^+/A^{2+} will exceed the ratio of the probabilities (4.5). It is well known that the appearance of intermediate resonances do not increase the probability of ionization more than by two or three orders of magnitude and thus cannot essentially change the above estimate of the ratio A^+/A^{2+}. If we take into consideration that in the reaction (4.3), the object of ionization is a positive ion but not a neutral atom, the true value of $\mathscr{W}_{11}$ will be much smaller than the one used in the estimate. This will increase the difference between the estimate and the experimental data. Thus $A^+/A^{2+} \gtrsim 10^{16}$, which strongly contradicts the observed value $A^+/A^{2+} \sim 10^3$.

Summarizing, one can state that the experimental data [5–7, 36–38] on the ratio of the outputs of singly and doubly charged ions cannot be explained by assuming that doubly charged ions are produced in a cascade process. Therefore in what follows we shall discuss the experimental data on the assumption that a direct process of nonlinear two-electron ionization of atoms (1) is realized, i.e., such a process when two electrons are detached from the atom simultaneously [37].

Further investigations of the two-electron process of ionization required clarification of the dependence of A$^+$ and A^{2+} ion output on the radiation frequency and polarization. At the present time such data are available for the case of Ba atom ionization only.

V. STUDY OF THE PROCESS OF NONLINEAR IONIZATION OF THE BARIUM ATOM

The first series of experiments, presented in Section IV, has shown that to have an insight into the process of doubly charged ion production, it is necessary to investigate the dependences on ion output on radiation frequency and polarization. The main results of frequency and polarization dependences of Ba$^+$ and Ba^{2+} ion output are presented below. Most of these results were published in Refs. [44–47].

A. Experiment

Different neodymium-glass lasers with a dispersion resonator and modulation of quality factor were used. The laser worked in the regime of generation of one transverse and many longitudinal modes. The laser made it possible to obtain an intense radiation at any of the frequencies in the following two overlapping intervals 9390–9460 and 9430–9500 cm^{-1}. The radiation spectrum width could vary from 10–0.3 cm^{-1}. The polarizer and the quarter-wave plate placed in a laser beam enabled the polarization to be changed from linear to circular with any arbitrary degree of ellipticity. The radiation was transformed into a second harmonic by a code relay crystal. The strength of the radiation field in the region of ion production was measured by a standard method [39], it ranged from 1.10^7–1.10^6 V cm^{-1} in different experiments. The absolute accuracy of the field strength measurement was $\pm 60\%$. The Ba atom density in the beam was about $\sim 10^{11}$ cm^{-3}.

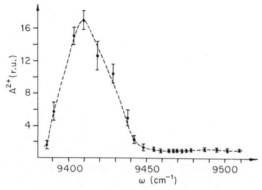

Fig. 2. Dependence of Ba^{2+} ion output on the laser radiation frequency (linear polarization).

B. Data on the Output of Ba^{2+} Ions

Dependence of the Ba^{2+} ion output on the frequency of linearly polarized laser radiation. Figure 2 presents the dependence of the output of Ba^{2+} on the frequency of linearly polarized laser radiation at the radiation spectrum width 3 cm^{-1} and radiation field strength 1.0×10^7 V cm^{-1}. One clear resonance is observed in Ba^{2+} ion output with the width of the order of 30 cm^{-1}, whose maximum corresponds to the frequency of 9410 cm^{-1}.

Dependence of the Ba^{2+} ion output in resonance on the laser radiation spectrum width. A large width of the resonance in Ba^{2+} ion output gave ground to begin studying the dependence of the width and form of this resonance on the exciting radiation spectrum width. The dependences of Ba^{2+} ion output in the frequency range from 9470–9465 cm^{-1} were measured using radiation with three different spectrum widths in the interval from 10.3 to 0.3 cm^{-1}. The results of experiments carried out at the field strength 1.0×10^7 V cm^{-1} are presented in Fig. 3. The absence of the dependence of the resonance form and width on the laser radiation spectrum width is clearly seen from this figure.

Dependence of Ba^{2+} ion output in resonance on the degree of radiation ellipticity. Figure 4 presents the dependence of Ba^{2+} ion output in a maximum resonance at a frequency of 9412 cm^{-1} on the degree of radiation ellipticity at the spectrum width equal to 3 cm^{-1}. The radiation field strength is 1.0×10^7 V cm^{-1}, $\theta = 0, 90°$ corresponds to linear radiation polarization; $\theta = 45°$ corresponds to circular polarization. The amplitude of Ba^{2+} ion output in circular radiation polarization was at the level of the limiting sensitivity of the device. The results of experiment show that Ba^{2+} ion output in resonance strongly depend on the degree of radiation ellipticity.

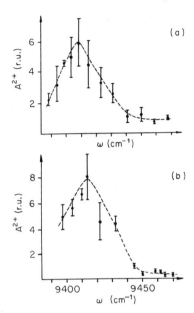

Fig. 3. Dependence of Ba^{2+} ion output in resonance on the laser radiation spectrum width. (a) $\Delta\omega \approx 10 \text{ cm}^{-1}$ and (b) $\Delta\omega \approx 0.3 \text{ cm}^{-1}$.

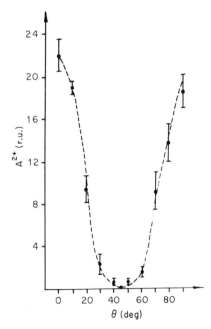

Fig. 4. Dependence of Ba^{2+} ion output in resonance on the degree of ellipticity of laser radiation. $\theta = 0.90°$ is linear polarization of radiation; $\theta = 45°$ is circular polarization.

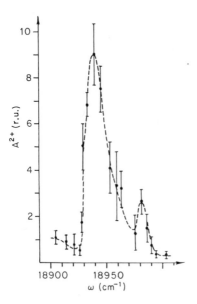

Fig. 5. Dependence of Ba^{2+} ion output on the frequency of the second harmonic of laser radiation.

Dependence of Ba^{2+} ion output on the frequency of the second harmonic of Nd-glass laser radiation. Figure 5 illustrates the dependence of Ba^{2+} ion output on the radiation frequency in the range from $18\,900–19\,000$ cm^{-1}. The radiation spectrum width was 4 cm^{-1}. The radiation field strength ranged from 1.5 to -2.0×10^6 V cm^{-1}. Two resonances are observed in this frequency range. The resonance width at the frequency $18\,940$ cm^{-1} was of the order of 25 cm^{-1}, and the resonance width at the frequency $18\,982$ cm^{-1} was difficult to estimate.

Additional information on the process of two-electron nonlinear ionization is provided by the experimental data on the output of Ba^+ ions.

C. Data on the Output of Ba^+ Ions

The dependence of the Ba^+ ion output on the frequency of linearly polarized laser radiation is shown in Fig. 6. The radiation spectrum width was 3 cm^{-1}, the radiation field strength was 6.0×10^6 V cm^{-1}.

The dependence of ion output on the frequency of linearly polarized second harmonic of laser radiation is given in Fig. 7. The width of the spectrum was 4 cm^{-1}; the field strength was 1.0×10^6 V cm^{-1}.

First, one should pay attention to the resonance at a frequency of 9410 cm^{-1}. This resonance coincides in frequency with the resonance in Ba^{2+} ion output (see Fig. 2). However, additional polarization measurements show (Fig. 8) that Ba^+ ion output in this resonance depends weakly

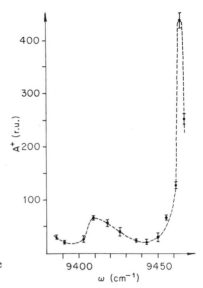

Fig. 6. Dependence of Ba$^+$ ion output on the laser radiation frequency (linear polarization).

on radiation polarization. We should recall that Ba^{2+} ion output in resonance at a frequency of 9410 cm^{-1} depends strongly on radiation polarization (Fig. 4). Taking into account this difference, it is difficult to suppose that one intermediate two-electron state in the barium atomic spectrum can be responsible for the appearance of the above-mentioned resonances.

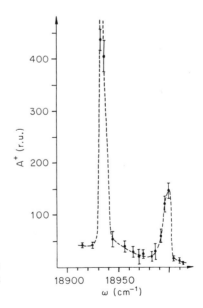

Fig. 7. Dependence of Ba$^+$ ion output on the frequency of the second harmonic of laser radiation (linear polarization).

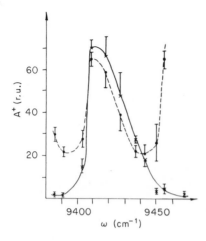

Fig. 8. Dependence of Ba$^+$ ion output on the radiation frequency in the resonance volume at a frequency of 9410 cm^{-1}; linearly polarized radiation (●); circularly polarized radiation (×).

Second, resonances in Ba$^+$ ion output at a frequency of 9462 cm^{-1} of the main laser radiation (Fig. 6) and at a frequency of 18 924 cm^{-1} of the second harmonic of the laser radiation (Fig. 7). As is seen from the above data, these resonances are observed at multiple frequencies. These resonances should be assumed to be due to excitation of one intermediate state in the barium atomic spectrum.

If we compare the resonance in Ba$^+$ ion output, observed at a frequency of 18 924 cm^{-1} (Fig. 7), and the resonance in Ba^{2+} ion output, observed at a frequency of 18 940 cm^{-1} (Fig. 5), the asymmetry of the second resonance may be supposed to result from the appearance of the first resonance.

Thus the experimental data on Ba$^+$ ion output testify to the possibility of the existence of correlation between Ba$^+$ and Ba^{2+} ion outputs.

D. Main Conclusions Concerning the Process of Ba^{2+} Ion Formation

The experimental data presented confirm the assumption of the existence of the process of two-electron nonlinear ionization.

Indeed at different radiation frequencies explicitly pronounced resonances in Ba$^+$ ion output are observed. Ba^{2+} ion output in resonance depends essentially on laser radiation polarization.

We would like to emphasize especially a large width of the observed resonances. So the half width of resonance maxima in ion output makes up 25–30 cm^{-1}. Remember that the experiment with a variable width of the laser radiation spectrum showed that the resonance width in the ion output is not connected with the laser spectral width (see Fig. 3).

Two physical reasons are known, which can in this case determine the resonance width, namely, ionization broadening of the resonance state and a shift of resonant states, which arises under the action of the radiation field [39, 48]. Estimates of perturbation of the Ba atomic spectrum, carried out in Section II.C, show that the resonance observed at the fundamental laser frequency for the field strength of the order of 10^7 V cm^{-1} can be, in particular, due to the shift of resonance levels. However, in the radiation field of the second harmonic of laser radiation with the strength of the order of 10^6 V cm^{-1} the level shifts are small. In this case one should assume the resonance width to be determined by the probability of transition of two electrons from a resonant state to a continuum. If one can use the concept of ionization probability per unit time, the ionization width of the order of 30 cm^{-1} corresponds to the ionization probability of the order of 10^{12} s^{-1}. However, at a duration of the laser radiation pulse of the order of 10^{-8} s and at such a probability, the transition to a continuum is saturated; this transition should be described in terms of the total probability.

Correlation between resonant frequencies for Ba$^+$ and Ba^{2+} ion outputs can be the result of a competition between different channels of reaction, which leads to Ba$^+$ and Ba^{2+} ion production under excitation of one and the same bound two-electron state in the Ba atomic spectrum. Such a competition can be due to the appearance of resonances in the spectrum of autoionization states. As follows from the data presented in Section II.B, resonances with autoionization states must occur practically at all radiation frequencies, when the atom is excited up to energies exceeding the first ionization potential. Therefore, the multiphoton transition of two electrons into a bound two-electron state is always followed by the one-photon transition into the autoionization state.

Let us now turn to resonances in Ba$^+$ ion output at multiple frequencies of 9462 cm^{-1} (Fig. 6) and 18 924 cm^{-1} (Fig. 7). Owing to a relative smallness of the strength of the radiation field of the second harmonic (1.10^6 V cm^{-1}) the observed resonance (Fig. 7) can be compared, and not without reason, with a nonperturbed spectrum of the barium atom. It is seen that this resonance can be due to a two-photon transition from the ground state of the barium atom $6s^2$ $^1S_0$ to a two-electron excited state $5d6d$ $^1D_2$. From the coincidence of the resonant frequencies, one should assume that the same state is excited also by four photons of the fundamental frequency of laser radiation. Note that at linear radiation polarization these transitions are allowed by selection rules for multiphoton transitions [2]. One should assume that, in this particular case, perturbation of resonant states in the barium atomic spectrum in the radiation field of the fundamental frequency is smaller than follows from the above general estimate (see Section II.C). If the above identification of resonant state is valid, this particular case is an example of reaction when relaxation turns out predominant; resonance is observed only in Ba$^+$ ion output. Note that the asymmetry in the resonance Ba^{2+} ion output, which is observed in this frequency range (Fig. 5) can be associated with competition of the reaction leading to the resonance Ba$^+$ ion output.

Analysis of experimental results, on the whole, agrees satisfactorily with this scheme of resonance nonlinear ionization of Ba atom—there appears first a multiphoton (four-photon on the fundamental frequency and two-photon on the second harmonic) resonance with a bound two-electron state, and then a one-photon resonance with autoionization state in the atomic spectrum. Further development of ionization process is determined by a relative value of partial widths of autoionization and two-electron ionization of a given state. Two-electron ionization, leading to Ba^{2+} ion formation, occurs here owing to cascade one-photon transitions in the spectrum of autoionization states and has the character of diffusion along the energy scale [49]. Autoionization leads to Ba$^+$ ion formation.

Note that the ideas about the process of doubly charged ion production developed here make it possible to explain why no doubly charged ions of Ca and Mg atoms were observed in the experiments described above [36, 37]. The reason may be a relatively large width of autoionization states of these light atoms as compared with that of Ba and Sr atoms (see Section II.B). Therefore, in excitation of autoionization states of Ca and Mg atoms, relaxation of these states turns out a determining channel of reaction; only singly charged ions are formed.

To summarize, we shall formulate the main conclusions drawn from the experimental data on Ba atom ionization, which seem to us to be certain:

Ba^{2+} ion output is of resonance character;
Ba^{2+} ion output depends essentially on radiation polarization;
resonance width in Ba^{2+} ion output is extremely large.

Different considerations and estimates are presented below that confirm the validity of the model of the two-electron ionization process formulated for the first time in Ref. [37], specified further on in Refs. [44, 49] and previously discussed in detail in this section.

VI. THE PROCESS OF TWO-ELECTRON NONLINEAR IONIZATION OF ALKALINE-EARTH ATOMS

Let us consider the model proposed from the viewpoint of its agreement with the known data on the processes arising in the interaction of atom with a strong light field.

Let us first turn to the first stage of the process—transition of two electrons from the ground state to the spectrum of autoionization states. In the general case, such a transition has a character of a direct multiphoton transition or a resonant transition provided that there appears resonance between the energy of several radiation quanta and the energy of transition from the ground to the bound two-electron state. A high density of the spectrum of autoionization states, as well as their strong broadening in the laser radiation field (see Section II.B), lead to the fact that practically at any radiation frequency, and as a result of absorption of several quanta, a corresponding autoionization state is excited. The appearance of intermediate resonance with a bound two-electron state increases the probability of transition of two electrons into the spectrum of autoionization states, as is always observed in the case of multiphoton transition at intermediate resonance [39]. It is just intermediate resonances with bound two-electron states that are responsible for the observed resonances in Ba^{2+} ion output. In the case when an intermediate resonance with a bound state is absent, the Ba^{2+} ion output depends weakly on the radiation frequency, as is always observed for a direct process of multiphoton ionization of atoms [39].

A multiphoton transition of two electrons into a bound two-electron or a low-lying autoionization state occurs as a result of absorption of approximately the same number of quanta, as in the ordinary process of one-electron multiphoton ionization of the atom. Indeed, in the case of ionization of the barium atom by the ruby laser radiation, bound two-electron states are excited as a result of absorption of four photons, and the absorption of five photons results in the excitation of autoionization states and in one-electron ionization. Since, following the proposed model, it is just multiphoton transition into a bound two-electron into an autoionization state that determines the probability of Ba^{2+} ion formation (see below), it becomes clear why the Ba^{2+} ion output differs little from the Ba^+ ion output.

Now let us turn to the transition of two electrons from a bound two-electron state into an autoionization state. The experimental data on resonances, presented above in Section V.B, make it possible to estimate the strength of oscillator for such a one-photon transition. We shall proceed from the resonance width $\simeq 30\,cm^{-1}$ observed at the field strength $\sim 10^6\,V\,cm^{-1}$. Using the known relations, (see, for example, [50]), the estimation gives the oscillator's strength value $f \sim 10^{-9}$. Unfortunately, there are neither experimental nor theoretical data to compare directly the above-mentioned value. Indirect information can be obtained, for example, from comparison of this value with the oscillator strength for one-photon transition from the ground state of the Ba atom to different autoionization states at the wavelength of about 1650 Å, measured in Ref. [18]. For final states with the principal quantum number $n = 13-16$ the value $f \sim 10^{-5}$ is obtained. It is apparent that in our case the oscillator strength must be smaller than this value. Indeed the transition takes place from a highly excited state; second, this state is two-electron and the wave functions of two excited electrons oscillate strongly and incoherently.

Thus our assumption that the width of the observed resonances in Ba^{2+} ion output is determined by the one-photon transition from a bound two-electron state into an autoionization state is in a reasonable agreement with the data available.

It is just at this stage of transition of two electrons into an autoionization state that there occurs competition between two channels of reaction. One channel is relaxation of the autoionization state. The other channel is a further excitation of two electrons along the spectrum of autoionization states. Relaxation results in Ba^+ ion formation and excitation to Ba^{2+} ion formation. Note that competition of relaxation is maximum just on the first resonance with autoionization states since the width of these states decreases rapidly as their energy increases (see Section II.B, Table III). It is obvious that competition of relaxation must depend essentially both on the exciting radiation frequency and on a specific structure of the atomic spectrum. From Table III it is seen that, for example, the widths of autoionization levels of the Ca atom are approximately greater by an order of magnitude more than those of the Ba atom. This, in particular, may be the reason why Ca^{2+} ions were not observed (see Section IV [36]).

Now let us turn to the process of the increase of energy of two electrons in the spectrum of autoionization states. The data on the density of autoionization states in the interval from the first to the second ionization potentials (see Section II.B) show that the assumption concerning the diffuse character of the electron energy increase [51] is rather well grounded. Indeed, the calculation of the spectrum of autoionization states of the barium atom [25] shows that all the transitions in the spectrum of autoionization states from the first to the second ionization potential are of the one-photon resonance character. In this case the probability of transition of two electrons from the initial autoionization state, lying near the first ionization potential, into the state with an energy corresponding to the second ionization potential, is the quantity inversely proportional to the time of electron diffusion along the spectrum of autoionization states. The diffusion time can be approximately estimated proceeding from the results of Ref. [51]. (Approximate character of this estimation is mainly connected with the fact that in [51] the transition along the spectrum of highly excited one-electron states was considered, whereas in this case we deal with transitions along the spectrum of two-electron states.) The approximate estimate shows that the diffusion proceeds for very small times, of the order of atomic times.

From this estimate it follows that for a two-electron ionization, it is practically sufficient to excite two electrons to the region of low-lying auto-ionization states whose energy is only a little higher than that of the first ionization potential. It is just this transition that determines all the main regularities of the two-electron ionization process—its probability, the dependence of the probability on frequency and on radiation polarization. Remember that in this case an essential role is played by the presence or absence of an intermediate resonance with a bound two-electron state.

In conclusion, we would like to note that the Ba^+ ions observed in experiment are produced not only as a result of relaxation of an excited two-electron autoionization state, but also owing to an ordinary one-electron multiphoton ionization process caused by transitions of one electron along the spectrum of one-electron states. Depending on frequency and radiation polarization, this process may have a direct or a resonance character. Turning to the experimental data on Ba^+ ion output, one should always take into consideration the possibility of their production both due to one-electron and two-electron transitions, which naturally complicates the interpretation.

VII. CONCLUSION

The proposed model of the process of two-electron ionization of atoms has been mainly discussed here using an example of the experimental data obtained for the barium atom. However, one should bear in mind that this model also explains the available data for other atoms. In our opinion one should expect that the model proposed will also, for the most part, be valid for other atoms whose spectra are similar to the barium atomic spectrum, in particular, for alkaline-earth atoms and lanthanides (with an account taken of the details of the structure of their spectra). However, we in no way expect that the proposed model of the formation of doubly charged ions in nonlinear ionization will be valid for other many-electron atoms, in particular, for noble gas atoms. The point is that there is a strong difference in the structure of atomic spectra, which has already has been mentioned above.

We should recall that many-electron structure of some atoms may have in nonlinear ionization also other manifestations not connected with the formation of doubly charged ions. In particular, many-electron effects can, in a number of cases, determine the probability of one-electron multiphoton ionization of atoms, just as is the case with photoionization.

In conclusion, we point out the main problems that now exist in the theoretical and experimental study of the process of nonlinear ionization of many-electron atoms.

N. B. DELONE, V. V. SURAN, AND B. A. ZON

Let us first turn to experiment. All the data presented above, even those obtained for the barium atom, are far from being complete. Therefore a natural task of the experiment is accumulation of data on frequency and polarization dependence of the formation of doubly charged ions of different atoms. It is of importance here that the experimental data should be obtained at the minimum field strength, which enables these data to be analyzed using the spectrum of a nonperturbed atom.

As concerns the theory, we should point out two circumstances. First, the development of traditional methods of the theory of nonstationary perturbations for the calculation of different elementary nonlinearly optical effects is necessary; but that would outstep the limits of one-electron approximation. First, we mean calculations of the dynamical polarizability and one-electron multiphoton ionization of many-electron atoms. In particular, calculations are necessary that would take into account the spectrum of two-electron bound and autoionization states.

Second, it is also necessary to realize a wide program of calculations and measurements of the spectra of autoionization states of many-electron atoms, alkaline-earth atoms and lanthanides in particular, to obtain data on perturbation of autoionization states by a laser radiation field.

Another question, purely theoretical, is the one concerning diffusion ionization with an account taken of two-electron transitions and transitions which proceed with great variation of the principal quantum number of the state. It is just for such cases that quantitative results on diffusion ionization are necessary.

Only the realization of such a wide program of developing the theory and accumulating the experimental data will enable the description of the nonlinear ionization process of many-electron atoms to reach the level attained by the study of the process of multiphoton ionization of alkaline atoms.

1. N. B. Delone and G. S. Voronov, *JETP Lett.* (*Engl. Transl.*) **1**, 42 (1965); *JETP* **50**, 78 (1966).
2. N. B. Delone and V. P. Krainov, *in* "Atom in a Strong Light Field." Atomizdat, Moscow, 1978 (in Russian); Springer-Verlag, Berlin and New York, 1983.
3. B. A. Zon, N. L. Manakov, and L. P. Rapoport, *in* "A Theory of Multiphoton Processes in Atoms." Atomizdat, Moscow, 1978 (in Russian).
4. Multiphoton atom ionization, *Tr. FIAN* **115** (1980) (Transactions of P. N. Lebedev Institute, in Russian).
5. V. V. Suran and I. P. Zapesochny, *JETP Lett.* (*Engl. Transl.*) **1**, 973 (1975).
6. I. S. Alexakhin, I. P. Zapesochny, and V. V. Suran, *Ukr. Fiz. Zh.* (*Russ. Ed.*) **21**, 1383, (1976).
7. V. V. Suran, *Opt. Spectrosc.* **41**, 901 (1976).
8. S. E. Frish, *in* "Optical Atom Spectra." Fizmatgiz, Moscow (1963) (in Russian).

9. C. E. Moore, "Atomic Energy Levels," Vols. 1–3. Washington, 1952.
10. M. Aymar, P. Camus, M. Dienlin, and C. Morrilon, *Phys. Rev. A.* **18**, 2173 (1978).
11. J. A. Armstrong and J. J. Wynne, *J. Opt. Soc. Am.* **69**, 211 (1979).
12. M. G. Kozlov, *in* "The Absorption Spectra of Metal Vapours in Vacuum Ultraviolet." Nauka, Moscow, 1981.
13. C. C. Gallagher, W. E. Cooke, and K. A. Safinya, *Laser Spectrosc., Proc. Int. Conf., 5th, Rottachoegerm,* p. 273 (1979).
14. J. J. Wygne and J. P. Hermann, *Opt. Lett.* **4**, 166 (1979).
15. D. J. Bradley, P. Ewart, V. J. Nickolas, and J. R. D. Shaw, *J. Phys. B.* **6**, 1594 (1973).
16. A. S. Nagvi, M. Y. Mirza, D. J. Semple, and W. W. Duley, *Opt. Commun.* **37**, 356 (1981).
17. V. L. Carter, R. D. Hudson, and E. L. Breig, *Phys. Rev. A.* **4**, 821 (1971).
18. R. D. Hudson, V. L. Carter, and P. Young, *Phys. Rev.* **180**, 77 (1969); *Phys. Rev. A.* **2**, 643 (1970).
19. W. R. S. Garton and K. Codling, *Proc. Phys. Soc., London* **75**, 87 (1960).
20. R. Ditchburn and R. Hudson, *Proc. R. Soc. London, Ser. A.* **256**, 53 (1960).
21. M. G. Kozlov and G. P. Startsev, *Opt. Spektrosk.* **28**, 14 (1970).
22. M. G. Kozlov and G. P. Startsev, *Opt. Spektrosk.* **28**, 1217 (1970).
23. E. P. Vidolova-Angelova, L. N. Ivanov, E. P. Ivanova, and V. S. Letokhov, *Izv. Akad. Nauk SSSR, Ser Fiz.* **45**, 2301 (1981).
24. S. A. Kotochigova, I. I. Tupitsyn. *All-Union* Conf. Coherent and Nonlinear Optics, 2nd, 1982. Abstract of paper.
25. P. Vogel, *Nucl., Instrum. Methods.* **110**, 241 (1973).
26. J. Tessman, A. Kahn, and W. Shockley, *Phys. Rev.* **92**, 890 (1953).
27. H. Schwartz, T. Miller, and B. Bederson, *Phys. Rev. A* **10**, 1924 (1974).
28. S. Chin, D. Feldmann, J. Krautwald, and K. Welge, *J. Phys. B.* **14**, 2353 (1981).
29. T. Carlson, *Phys. Rev.* **156**, 142 (1967).
30. R. Cairus, H. Harrison, and R. Scoen, *Phys. Rev.* **183**, 52 (1969).
31. T. Carlson and M. Krause, *Phys. Rev. A* **137A**, 1655 (1965); *ibid.* **140**, A1057 (1965).
32. A. M. Dykhne and G. L. Yudin, *Usp. Fiz. Nauk.* **125**, 377 (1978).
33. G. Wight and M. Van der Wiel, *J. Phys. B.* **9**, 1319 (1976).
34. J. Connerade and M. Martin, *J. Phys. B.* **13**, L373 (1980).
35. M. Amusia, *Comments At. Mol. Phys.* **10**, 155 (1981).
36. V. V. Suran, *Ukr. Fiz. Zh. (Russ. Ed.)* **22**, 2055 (1977).
37. N. S. Alexakhin, N. B. Delone, I. P. Zapesochny, and V. V. Suran, *Zh. Eksp. Teor. Fiz.* **76**, 887 (1979) [*Sov. Phys. JETP,* **49**, 447 (1979)].
38. S. Chin, A. von Hellfeld, J. Krautwald, D. Feldmann, and K. Welge, *Abstr. Contr. Pap. 2nd Int. Conf. Multiphoton Process,* B-12, Budapest (1980).
39. N. B. Delone, *Usp. Fiz. Nauk* **115**, 361 (1975).
40. T. U. Arslanbekov, V. A. Grinchuk, N. B. Delone, and K. B. Petrosyan, *Kr. Soob. Po. Fizike* **10**, 33 (1975).
41. N. Manakov, V. Ovsyannikov, M. Preobrazhensky, and L. Rapoport, *J. Phys. B.* **11**, 245 (1978).
42. D. T. Alimov and N. B. Delone *JETP* **70**, 29 (1976).
43. N. B. Delone, N. L. Manakov, M. A. Preobrazhensky, and L. P. Papoport *Zh. Eksp. Theor. Fiz.* **70**, 1234 (1976).
44. N. B. Delone, I. P. Zapesochny, B. A. Zon, and V. V. Suran, *Izv. Akad. Nauk SSSR, Ser. Fiz.* **45**, 1081 (1981).
45. I. I. Bondar', N. B. Delone, I. P. Zapesochny, and V. V. Suran, *JETP Lett. (Engl. Transl.)* **7**, 243 (1981).
46. I. Bondar', N. Delone, B. Zon, and V. Suran, *Opt. Commun.* **40**, 268 (1982).
47. I. I. Bondar', N. B. Delone, I. P. Zapesochny, and V. V. Suran, *Preprint FIAN* No. 111 (1982) (in Russian).

48. N. B. Delone, and M. V. Fedorov, *Tr. FIAN* **115**, 42 (1980) (Transactions of P. N. Lebedev Institute, in Russian).
49. N. B. Delone, B. A. Zon, V. P. Krainov, and M. A. Preobrazhensky, *JETP Lett.* **30**, (*Engl. Transl.*) 260 (1979).
50. I. I. Sobelman, "Introduction to a Theory of Atom Spectra." Nauka, Moscow, 1977 (in Russian).
51. N. B. Delone, B. A. Zon, and V. P. Krainov, *Zh. Eksp. Theor. Fiz.* **75**, 445 (1978), [*Sov. Phys. JETP* **48**, 223 (1978).]

SUPPLEMENTARY REFERENCES

Note added in proof: Since this chapter was written the following papers have been published, devoted to different problems pertaining to the subject of this chapter. These papers confirm in the whole the basic conclusions made in the chapter.

1. D. Feldmann, J. Krautwald, S. Chin, A. von Hellfeld, and K. Welge, *J. Phys. B.* **15**, 1663 (1982).
2. D. Feldmann and K. Welge, *J. Phys. B.* **15**, 1651 (1982).
3. D. Feldmann, J. Krautwald, and K. Welge, *J. Phys. B.* **15**, L529 (1982).
4. A. L'Huillier, L. Lompre, G. Mainfrag, and C. Manus, *Phys. Rev. Lett.* **48**, 1814 (1982).
5. M. Preobrazhensky, *Zh. Eksp. Theor. Fiz.* **83**, 1985 (1982).
6. N. Delone, V. Krainov, and D. Shepelyansky, *Usp. Fiz. Nauk* **140**, 355 (1983).
7. S. Kotochigova and I. Tupitsyn, *Izv. Akad. Nauk SSSR, Ser. Fiz.* **47**, 1142 (1983).

Index

A

Above-threshold ionization, 133–153
 definition, 134
 dispersive methods, in electron spectroscopy, 142–143
 versus inverse bremsstrahlung, 151
 nondispersive methods, in electron
  spectroscopy, 142
 nonperturbative treatment, 136–139
 numerical calculations, 140–141
 theory, 135–141
 Zernik calculation, 140
Above-threshold ionization experiments,
  141–151
 conditions, 143–144
 electron spectroscopy, 142–143
 experimental results, 145–151
 intensity dependence of peak amplitudes, 149–150
 use of lasers, 143–145
Above-threshold ionization probabilities,
  perturbative calculations, 139
Ac-Stark effect, in angular distribution of
  photoelectrons, 127–129
Ac-Stark splitting
 autoionizing resonance, 210
 multiphoton autoionization, 211–212
Alkaline earth atoms
 absorption, multiphoton, 191
 atomic spectrum perturbation, 243–245
 autoionization states, 240–243

bound states, two-electron, 238–240
doubly charged ion formation, 248
ionization, multiphoton, 236
nonlinear ionization, two-electron, 258–
  261
specific features in spectra, 237–245
Angular distribution of photoelectrons
 in above-threshold ionization experiments, 148–149
 autoionizing states, 123–125
 basic formulas, 112–114
 bound–free matrix and scattering
  phases, 120–122
 experiments, 115–129
 formalism, 104
 ionization
  multiphoton, 111–129
  off-resonant multiphoton, 129
  three-photon, of sodium atoms; via
   resonant intermediate states, 115–
   119
  two-photon, of cesium atom, 27–31,
   120
  two-photon, of sodium atoms, via
   resonant intermediate states, 115–
   119
 and perturbation, by configuration mixing, 122–123
 polar diagrams, 116
 quantum interference effects, 125–127
 and Stark effect, 127–128
 theory, 112–114